Our Human Current:

Stories about Complexity, Systems, and Mentorship

Our Human Current:

Stories about Complexity, Systems, and Mentorship

Angela Cross and Haley Campbell-Gross

HPT Development, LLC
Austin, Texas

HPT Development, LLC
925 S. Capital of Texas Hwy, Suite B245
Austin, Texas 78746

human-current.com

This book contains material previously published by HPT Development, LLC, on the *Human Current* platform at human-current.com. Grateful acknowledgement is made to HPT Development, LLC; co-hosts Haley Campbell-Gross, Angela Cross, Stacy Hale; and podcast guests and guest bloggers for permission to reprint material from *Human Current*.

For ordering information, visit human-current.com.

ISBN 978-0-578-50983-9

Cover artwork by JK Rofling at jkrofling.com.

Edited by Cynthia Scheiderer.

Book design by Dorene Taylor.

For our mentor Douglas Drane, our podcast guests, and our listeners.

With gratitude.

Table of Contents

our
Preface: About ~~the~~ ^ *Human Current*

"Welcome to *Human Current*,
where we have casual conversations about complexity
in work, life, and all the things we do.
It's time to get curious and curiouser."
Human Current podcast introduction

The phrase "write to your audience" has always made us cringe a little. You can't know what mental models your listeners or readers are running. You can't control what circumstances they're experiencing when they read your work. What you *can* do is hope that by being as authentic and curious as you can be, you will connect with others, and wherever *you* are, and wherever *they* are, is where you both need to be to have a great conversation. Nature doesn't need our help to move information through its networks. The world is always changing. But wherever our shared curiosity has directed its movement from mind to matter and back again, our human current can be found.

This is the spirit in which we've written this book. *Human Current* began as a passion project of our mentor, Douglas (Doug) Drane, to spread awareness of complex systems science and systems thinking. With Doug, Angela (Angie) Cross, Stacy Hale, and Lindsay Patterson founded *Human Current* in 2015. After the departure of Patterson and Hale, Cross welcomed Haley Campbell-Gross as her new co-host. Together we continued *Human Current* until, over the course of four years, we had published 125 podcast episodes featuring a wide range of experts, practitioners, scientists, and enthusiasts from the complexity science community. Our podcast journey included stories, research, anecdotes, and advice from our amazing guests, including: Stephen Wolfram, a founding father of complexity science; Yaneer Bar-Yam, president of the New England Complex Systems Institute; Dave Snowden, founder of Cognitive Edge; Jean Boulton, co-author of *Embracing Complexity*; Melanie Mitchell, author and professor at the Santa Fe Institute; César Hidalgo, author and professor at Massachusetts Institute of Technology; Margaret Wheatley, best-selling author and chaos theory practitioner; Dean Radin, chief scientist at the Institute of Noetic Sciences; Albert-László Barabási, author and professor of Network Science; and so many, many more!

When our podcast began, we recognized that we were facilitators of a collective journey, so we thought of *Human Current* as really *our* human current, a place for complexity and systems thinkers from all walks of life to join together in exploratory conversations. This book is a reflection of our journey together and a celebration of the diversity of stories we co-created. Our intention setting out to write this book was two-fold: to honor Doug's mentorship, and to include, even briefly, every single one of the 125 episodes we produced. The episodes bring you stories and quotes from many different voices—each providing a unique perspective and insight from which we can learn. Some of the insights shared in this book are specific to certain systems—like the human brain, politics, the environment, or company culture—while other insights are more universal, helping us understand how and why systems shape our world.

Purposefully and with loving intent, we designed this book to take you on a 365-day journey through stories of our *Human Current* and Doug's mentorship. We invite you to start anywhere and move freely around as you read this book; after all, our stories are not linear or sequential, so neither should your experience be with this book. In the spirit of Doug's love for books, scribbling in the margins of books, learning from books, re-reading the same books over and over again, we encourage you to write in the margins, fold over the page corners, and share your ah-ha moments with others. Our greatest hope is that these brief reflections will inspire you to learn more from our guests. For example, listen to the full podcast episode, explore a guest's website, watch a *TED Talk* they gave, or read a book or article they published. Maybe even sip a glass a wine while you explore the vast world of complexity and all our guests have to offer.

Cheers!

Acknowledgements

We have referred to Douglas (Doug) Drane as our friend, mentor, and our raison d'être for our work at *Human Current*. We continue to be in awe of Doug's life experience. Doug, who was one of five siblings, was born in 1940 and raised in Jackson, Mississippi where he attended St. Joseph's High School and played football. Even though Doug went on to attend elite schools such as Notre Dame University, Princeton Graduate School, Harvard Business School, and the Massachusetts Institute of Technology, he attests to his most profound learning being from his earliest years of employment. When Doug was twelve years old, his older brother Tupper landed him a job at Parkin's Pharmacy, and then at fifteen years old Doug worked at Baker's Shoe Store. These early experiences taught him valuable sales skills and impressed on him the importance of listening carefully to exactly what a customer needs. He kept this focus on the customer's experience throughout his business career.

In the 1970s, Doug and brothers Charlie and Richard Ying founded Atex, a company that revolutionized the publishing industry. Doug then spent more than 50 years studying and designing ventures around how to build high-performance teams that self-organize into successful companies. One of these ventures was HPT Development, home of *Human Current*. His life's work is a system for the optimization of each individual employee and the teams they're on to reach maximum effectiveness. His complexity model presents an understanding of business as a complex adaptive system and is based on decades of collaboration with top minds across many disciplines to understand how small teams of aligned, high-performing individuals can change the workplace for everyone. Doug wholeheartedly believed that if more people understood complexity, the world would be a better place, and it was this belief that sparked *Human Current*.

We are grateful to Doug for his vision and passion, his generous funding of *Human Current*, and his continuous encouragement and mentorship. We also extend a big thank you to our other *Human Current* co-founders, Stacy Hale and Lindsay Patterson. Their enthusiasm for the vision and commitment to the podcast helped build the foundation for *Human Current's* success.

Hale was an original co-host for *Human Current*. She has a background as a brand copywriter and journalist, and describes herself as a curious person who is obsessed with complexity theory and the study of human potential and consciousness. While recording casual conversations with co-host Angie Cross, her go-to wine was Prosecco. Hale, who is now a content director at IBM, provided most of the written content on the

Human Current blog during *Human Current's* first year, excerpts from which appear throughout this book.

Patterson, who has a background in public radio and an obsession with podcasts, was the behind-the-scenes producer and helped develop the *Human Current* podcast, put together the earlier shows, and performed an editorial role. Before co-founding *Human Current*, Patterson started *Tumble*, a science podcast for kids which she still co-hosts and produces.

Thanks to Winnie Shivers with Purple Lemon Co. for being an avid *Human Current* listener and transcribing all of the podcast episodes so they could be available online at human-current.com. Thank you to Noah Stiles with Stiles Sound Services for providing exceptional audio editing services, and to Logan Presley with Colossus Sounds Studio for producing *Human Current* videos.

A special thank you to our editor, Cynthia Scheiderer, for sharing her love, brilliance, and countless hours during the editing process. Scheiderer not only provided a critical service, she went above and beyond to ensure this book was successfully completed. Thank you to our graphic designer, Dorene Taylor, with Digital Fusion Group. Taylor graciously and patiently coached us through the graphic design and printing process so we could make the best formatting, design, and printing decisions.

To the more than 125 guests who spent time with us and shared their wisdom, stories, and hearts: this book would not exist without you. We appreciate you.

Thank you to our HPT Development work family and our home families—especially our amazing husbands for their love and support of our work at *Human Current* and while writing this book.

Finally, to our global audience of more than 150,000 listeners: thank you all for listening, for your feedback, and for being a part of our journey.

If more people understood complexity,
the world would be a better place.

Douglas Drane

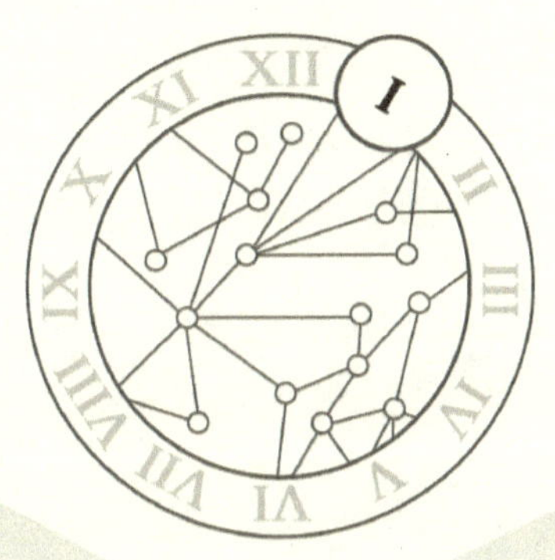

JANUARY

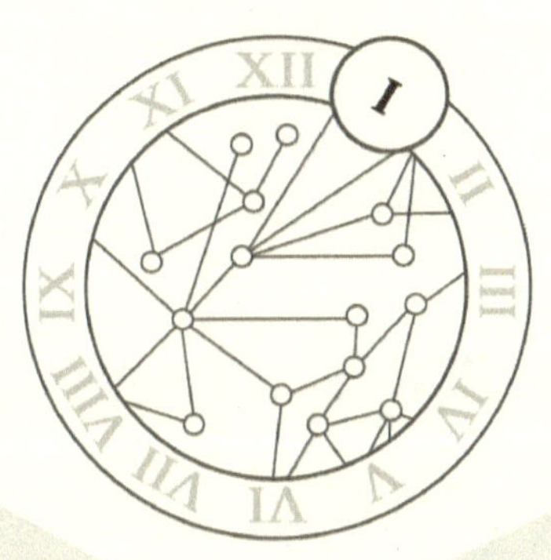

JANUARY

1

The Complexity Model

Douglas Drane, mentor and co-founder of Human Current.

The *Human Current* podcast explores complexity science and systems thinking in the workplace and beyond. *Human Current* was established as a passion project of our supporter, cheerleader, mentor, and mastermind, Douglas Drane. He is the reason our team existed, and it was his love for complexity that we shared with the world every day.

Doug is a lifelong entrepreneur and the founder of HPT Development, the home of *Human Current*. His life's work has been focused on his complexity model for business development, which understands business as a complex adaptive system. The model was based on decades of collaboration with brilliant minds across many disciplines to understand how small teams of aligned, high-performance individuals can change the workplace for everyone. In countless inspiring conversations with Doug, he would talk about his belief that the world would be a better place if more people were complexity thinkers. This became the spark that brought *Human Current* to life.

Four amazing years since starting *Human Current*, we continue to learn about complexity science and grow as complexity thinkers. Our curiosity to further understand and apply complexity thinking at work and in other areas of our lives led us to other opportunities, such as going to graduate school. Our mindset and understanding of the world and its complex adaptive systems shifted. With new opportunities we experienced "metanoia": a journey of changing our minds, hearts, selves, and our way of life.

The lessons we have learned at *Human Current* have fostered our complexity thinking. As a result, we became better co-hosts, employees, friends, and humans. Doug was right. The more complexity thinkers there are, the better the world will be.

Thank you, Doug!

We build collective intelligence by living the questions together, by understanding what answers are in the room. And just by listening to each other, we soften and we experience our interconnectedness and our common humanity, our human current. We are part of this river of life flowing into a future. We are co-creating that future by the way we tell stories about it and the way we participate in it. We all shape the future every day, just a little bit, and that makes us enormously powerful and enormously humble.

Daniel Wahl
Episode 122 Design-Led System Transformation

3

The *Human Current* Journey

Doug's one and only mission for us was to "change the world," which might seem like an extraordinary task, but in reality, it was a simple one. We just had to realize that changing the world starts with one person and one idea. For our story, that person was Doug and the idea was his complexity model.

Complexity is what makes life interesting and beautiful, and it is our job to recognize and respect how perfectly the universe operates without our interference. In fact, it is our desire to control that is our downfall. Ironically, the more we try to enforce our own structures, rules, and theories to the systems around us, the less control we have over them. In an article for *Harvard Business Review*, "Turning 60: The Twelve Most Important Lessons I've Learned So Far," Tony Schwartz said, "Let go of certainty. The opposite isn't uncertainty. It's openness, curiosity, and a willingness to embrace paradox, rather than choose up sides."

We can learn so much just from closely observing the world and the systems around us. Plus, what we learn can be applied to all areas of our life. It's that simple. Well, it's supposed to be simple! Yet it's a foreign concept to most of us. For that reason, we made it our mission to change the world with one word: curiosity.

Our journey began and continued with a spirit of curiosity. We used curiosity to fuel our conversations about complexity, to inspire others to share their stories, and to find the hidden connections that help us all understand the world a little better.

Over the years, we explored the idea that we can use a complexity lens to examine everything in life, and we tested that theory by asking tons of questions of brilliant people from many different fields. Some of these fields included: environment, politics, health care, education, design, company culture, emergency planning, and social justice and equity.

At *Human Current*, we facilitated open-minded conversations with more than 120 of these thought leaders from around the world to explore many applications of complexity in work and life. Their stories and our continuous reflection helped us trace the connections and interrelationships within and between different systems.

It has been an exciting journey. Thank you to Doug, our audience, and all of our guests who contributed to our mission.

4

What Is a Complex Adaptive System?

Just what is a complex adaptive system? Discussing the different perspectives and definitions thinkers brought to the topic was one of the joys of hosting the *Human Current* podcast. For a starting place, we like the definition offered by Lee G. Bolman and Terrence E. Deal in *Reframing Organizations: Artistry, Choice, and Leadership*. They write that a complex adaptive system is "a network, a set of loosely connected units, each pursuing its own agenda in response to local conditions." A key feature of complex adaptive systems is the absence of strong central control, which Bolman and Deal call "a virtue because local units can adapt very quickly to new developments, and because the loss of any one outpost does little damage to the whole."

5

Why Complexity?

A complex adaptive system is the product of many relationships that evolve together in response to their interactions with each other. Whether you're at work, at home with your family, or walking around in nature, *you* are part of many evolving complex systems.

You can't eliminate complexity. Life has too many unknowns and connections to eliminate complexity. Rather than fighting complexity, to get in tune with a complex system's behavior, look for "attractors." These are places in the system where patterns coalesce. Once relationships self-organize into patterns, these patterns cannot be erased. In complexity science this is called "history dependence." Like neural networks, connections cannot be undone, only replaced. Understanding patterns of interrelationship within the system can help us detect signals in the noise—we call this "coherence"—and from there we can make more sense of it.

Denying complexity, misunderstanding it, or trying to reduce it into simple causal chains can lead to taking shortsighted measures with disastrous results. For example, when people artificially introduce predation to "rebalance" an ecosystem, they can end up with endangered or eliminated native populations and starving predators. Unfortunately, many of our current business, economic, and social practices are focused on that kind of reductionist thinking: the endless search for a single cause and effect.

Complexity thinking is not about focusing on how complicated things are. It is about looking at all the relationships, dynamics, uncertainty, and unexpected or emergent behavior to detect patterns, find coherence, and measure the effects of our interventions.

6

Complexity Simply Is

We loved asking experts about complexity and getting . . . well, complex! We believe that conversations about nonlinear systems would—and should—be held everywhere and with anyone at any time. If you are part of a human organization that evolves—whether it's a company, club, or family—you are part of a complex adaptive system. If you have a body, you are a complex adaptive system.

Complexity arises when multiple forces or "agents" within a system interact on a local level, with feedback rippling through networks to produce emergent behavior and patterns. In other words, small shifts can lead to big changes. The results may not reflect the energy or effort involved in the interactions. The whole may be greater than—or more subtle than—the sum of its parts. Yet as they evolve, complex systems can exhibit timeless, unanticipated symmetry, as if the patterns were there in blueprint from the beginning. And isn't that what makes life interesting and beautiful?

7

Nodes in the Human System

Once you look at life through a systems lens, there's no going back—not even for fear of your own ignorance. As Denise Easton told us in Episode 117 Influencing Pattern-Based Change, "Complexity thinking is not only a mindset, it's a way of being. It just becomes embedded in your DNA."

It's like relearning the world in color, stepping into Oz from out of a Kansas dust storm of black and white. That was our purpose at *Human Current*. We aren't scientists or data analysts. We are just nodes in the human system who were seeing the world in a new way and wanted to talk about complexity with others who have pieces of the pattern.

Complexity runs like a current through our lives and nature without respect for the boundaries and mental borders we create. We hoped that by gaining a better understanding of how complexity works, we could relax a little more with less need for control in the sandboxes in which we play. We wanted to share what we learned because we believed thinking this way would make us smarter, happier, and more conscious. After four years, we can say: that's exactly what happened.

8

And So *Human Current* Began

The following is an excerpt from Episode 000 Welcome to *Human Current*. In this first episode the founding co-hosts, Stacy Hale and Angie Cross, discussed why a complexity podcast is important.

Hale: I love complexity because I love patterns. I believe that everything—man, machine, animal, mineral, vegetable—is part of an absolute, and when you can see patterns, synchronicities, and systems at work and at play in that absolute, it just sounds like music. It's just beautiful.

Cross: I've been exploring complexity and practicing my systems thinking lens for a while now. I love seeing the patterns and having those ah-ha moments where I start to recognize complexity in very simple everyday situations. And what I found is that the more I see with this lens, the harder it is to turn off. It's always on. Do you experience something similar?

Hale: Yeah, once you begin to see things in terms of systems and cycles and feedback, you start to see a lot of similarities across the different spheres of life and experience. It's like this woven tapestry that's self-similar, whether you go all the way down to the cellular level, or all the way up to the cosmological level. And you see it in so many disciplines. Personally, I love when philosophy and science meet at a crossroads.

Cross: Complexity thinking is so much about science *and* it's not at all about science.

Hale: It's about observation. . . . We call it natural curiosity and I think it's something that everyone has. You just have to find what you're most curious about and then look outward from there and see the relationship. I think that complexity is about interrelationships and recognizing it in all things. We don't think about the way our heart beats, but it still beats anyway. I think that systems are always at work everywhere and it's just a matter of turning our attention to them to see the patterns and the rhythm.

Cross: It's like we're not exploring the right or wrong in an interrelationship or the pattern. Rather, it's about growing appreciation for what exists and seeing it from different perspectives and different lenses. And maybe having that greater awareness and appreciation can help change the world.

Hale: Yeah, when you zoom out a little bit, you may see that there is a larger pattern at work. When something from one vantage point looks like chaos, from a wider vantage point or from a zoomed-in vantage point you might see a pattern. It's what we're looking for when we try to solve for pi. We think that if we zoom out far enough we will eventually see a pattern. Complexity is the opposite of linear, reductionist thinking. It acknowledges that all the individual units in a system interact and from those interactions new patterns emerge, but the system is always adapting, so it's always changing and so are its patterns.

Cross: We can apply that idea to anything in life and any situation in life, and I'm looking forward to exploring, because we're not experts! We're explorers and we want to be seen as this hub of complexity and systems thinkers to bring those experts to the table from all walks of life.

Hale: If you work in complexity, no matter what your discipline or field, we want to talk to you. We think it's a conversation that everybody should be having.

9

Patterns become a consistent construct that help give the form and function to the dynamics in an organization. So you can talk about an ecosystem pattern and they're really deeply embedded—you almost don't even recognize them—but if you really dive down and try to unbraid what they are, you have history, you have culture, and you work within a particular context. And those are really important.

Denise Easton
Episode 117 Influencing Pattern-Based Change

10

Understanding Consciousness Everywhere It Exists

A koan. Love. An unsolvable puzzle that looked easy at first. Like these mysteries, complexity hints at the ultimate mystery, calling us ever closer into deeper layers of understanding. We can learn about ourselves by looking at the complex world that produced us, at the complex things that we produce—and most of all, at the relationship between these things.

In nature, the flocking of birds reflects one important behavioral characteristic of complex adaptive systems. A flock of birds can appear to move in unison, shifting direction dramatically as if the flock shared one mind or was led by a single bird. In fact, the flock has no real leader or mind-melding abilities. What's actually happening is that one bird on the edge turns, prompted perhaps by a change in the wind; the bird next to it follows suit, going with the flow and avoiding collision; then the two birds next to that one shift their trajectory; and so on. The bird-to-bird feedback is so immediately integrated that the flock appears to move as one consciousness.

You can see how this translates to human social networks—especially cultures. The relationships between the agents of a complex system are where the real pattern-forming action happens, except that in complex systems there aren't always easy causal explanations to be found just by taking the system apart and looking at its components. To borrow an analogy from Robert Sapolsky, professor at Stanford University, imagine trying to take apart a cloud to understand how the weather works. And yet complex adaptive systems, however diverse their origin, have a fascinating number of characteristics in common—enough to form a discipline as complex as the subject of its study.

11

Complexity Is #NotBoring

Since relationships are constantly shifting and evolving in quality, context, and frequency, complexity guarantees that the system will eventually produce unexpected results, even (and especially) when small constraints are introduced. That guarantees that complexity is never boring!

To some, talking about complexity may appear to be so much intellectual frolicking in a field that's only recently been seeded. However, curious minds across disciplines including biology, climatology, economics, genetics, sociology, medicine, physics, mathematics, and many others have been talking about this for a while, just maybe not in the same room. What these people are really geeking out about is the possibility of understanding how the universe works once you can zoom far enough in and out to see patterns and relationships.

Complexity is increasingly getting some attention in the business world. For example, companies have experimented with holocracy, an approach to decentralizing leadership and promoting self-organization and self-management in business. Tony Hsieh, chief executive officer of Zappos, made news in 2013 when the company implemented holocracy, eliminating traditional manager roles. Holocracy aims to be a tool for navigating complexity in companies, both as a product and as a set of practices that is becoming a movement. And that's just the beginning of an intellectual crossover that promises to be an exciting—and definitely not boring—shift in business.

12

Why Complexity Is the New Simplicity

At *Human Current*, we often heard questions like: Why complexity? Why not focus on simplicity instead? We took on these questions in Episode 001 Why Complexity?

Systems are an often invisible fact of life and nature. Think ecosystems, cultures, and human relationships. It's hard to break these things down and understand them by examining their parts; however, from a systems perspective, you can zoom out and see for miles. Complexity science observes the relationships between the parts, and the outcomes they produce, in order to understand the ways the whole system is changing and organizing itself.

13

Napkins and Envelopes

"We can introduce loose frameworks and processes that enable people to be more adaptive." Stacy Hale arrived at this insight for Episode 015 Why Complexity? A Thursday Throwback Reflection. She had been reflecting on a keynote address at the Lean, Agile & Scrum Conference 2013 by Dave Snowden where he said, "Any framework or model that can't be drawn on the back of a table napkin from memory has little utility."

This is also something Doug talked with us about. He called it envelope planning, the action of quickly and efficiently writing down ideas in order to get a framework or strategy out quickly without getting caught in formal and rigid business plans.

We get in trouble when we start mistaking the model for reality. It's the concept that all models are false, but some are useful. A plan or design that can be mapped out on an envelope or napkin can help an organization stay agile. In other words, as in the aphorism often attributed to Albert Einstein, "Everything should be as simple as it can be, but not simpler."

14

Feedback and Change

When something changes and evolves over time, we call it dynamic. A city is a great example of a dynamic system, because feedback between the city and the people who interact with it form patterns of interrelationship that continually change its shape and character. Feedback shapes how change happens. We call this a complex adaptive system because there are many agents—factors that affect the system—that influence both *it* and *each other*. Changes in the system (the city) and its parts (the people) cause both to continually adapt.

Making Sense of Human Systems

We interviewed Dave Snowden, founder and chief scientific officer of Cognitive Edge, for Episode 044 Making Sense of Human Systems. Snowden described his work with complexity and how it applies to decision-making. He said, "There are huge additional levels of complexity when you deal with human systems, so what we are doing is to combine complex adaptive systems theory with cognitive neuroscience, psychology, and philosophy to create a new theory of what systems are about."

Snowden also introduced us to "anthro-complexity," a term he coined to represent a new approach to human systems based on natural science. Snowden said, "Human systems are fundamentally organic entities." Therefore, it doesn't make sense to study them from a mechanical framework. "Human systems are complex ecologies rather than complicated machines. For the last three or four decades we've been using a mechanical metaphor for what is fundamentally an organic entity."

16

Learning to See Complexity

If you are a systems thinker, you may not be able to recall a time when you couldn't see systems. Continuing to learn complexity vocabulary and concepts, and being able to wrap words and meaning around our experiences, has increased our confidence, even if we are still not completely fluent in "complexity speak." There are many resources that have helped us grow as complexity and system thinkers. Here are a few we want to share with you:

- Listen to complexity-related podcasts. How about starting with *Human Current*, the complexity podcast?
- Watch systems thinking and complexity-related *TED Talks*.
- Check out the courses at Systems Academy, systemsacademy.io.
- Learn new concepts and skills at the New England Complex Systems Institute, necsi.edu.
- Explore the Waters Center for Systems Thinking, waterscenterst.org.
- Read books about systems and complexity thinking. You can find a list of what's on our Bookshelf at the back of this book.

17

Why Worry About Complexity?

If we are asked, "Why worry about complexity when there are starving people in the world?" We respond, "Because economies, agriculture, and politics are complex systems."

Complexity is not some abstract academic concept that doesn't apply close to home. Complexity in the workplace, for example, is something that goes beyond common management best practices to look at the emergence of unintended outcomes of human interactions (i.e. culture, innovation, market perception) and how they affect and co-evolve with the organization's brand, customers, processes, level of bureaucracy, and informal networks. Complexity describes all that goes on behind the scenes, the entire ecosystem, that makes or breaks a company. One type of complexity strategy would be to plan for adaptability and emergence as opposed to managing to avoid failure. Companies can do this by creating loose frameworks that appropriately incorporate failure and flexibility so the system bends instead of breaks.

By understanding complex systems we shift from a measure-and-control mentality to a place of curiosity and creativity. When we know that the smallest action taken at the right time can make a big difference, we can explore change with less effort and watch how our actions ripple through the system. Dave Snowden calls it "a matter of guiding and directing flow." When we have the sort of flow that we can guide and direct, we work happier.

18

When Is It Right to Trust the Process?

Whenever you have a group of people interacting over time, you have a complex adaptive system. You can have clear hierarchies and role requirements, but what you can't easily quantify is how powerful a role emergence will play in how projects and people take shape when they are all co-evolving. This was especially true at *Human Current*, where we ran a flat organization. It was beautiful, really. Together we shaped the work, and the work changed us.

There are a lot of factors that have agency to cause the changing shape of any system. The most obvious agent for making change or progress is an idea. Here's the thing about ideas and how much power they have: when we use our past experiences to inform our judgments about the present, we're using a mental model—a form of idea—that gives us a structure to work with, because we need to see patterns in order to make meaning.

The interesting thing is, we all run different mental models based on different value systems and experiences, so no one is interpreting the present the same way. This really does make things interesting! It means potentially endless interpretations of problems and solutions. How do we come to consensus? Maybe awareness of the mental models we and others are running is a good place to start.

In decision-making in our teams, how much do we rely on mapping scenarios, working with existing structures, and when can we allow a little chaos to intervene? If innovation and sustainability are the goals, what's the right formula for trusting the process?

That's not a rhetorical question, by the way. We hope you will answer it for yourself!

19

The Science and Philosophy of Complexity

We had a discussion with Carlos Gershenson, research professor at the Self-Organizing Systems Lab at the Universidad Nacional Autónoma de México, for Episode 088 The Science and Philosophy of Complexity. Gershenson discussed findings from his book, *Complexity: 5 Questions*, which compiled interview-style contributions from leading figures in the field of complexity. He said, "I like to promote knowledge related to complex systems because I think to have some practical impact people need to know what science is offering to solve real-world problems."

Gershenson also shared his own perspectives on the past, present, and future of complexity science, as well as how philosophy plays a role in the emergence of science. He said, "As new generations become familiar with these ideas, I think there will be a change in our worldview and we will embrace this lack of predictability. And it will be better because, instead of trying to ignore the limitations that we have for prediction, we will be designing our systems and our societies so they can better cope with this lack of predictability by adapting and being more resilient."

20

Complicated Versus Complex

Complicated versus complex. This was one of our early big light bulb moments. A process or situation in a *complicated* system has a solution, even if it is puzzling and involves many steps. That's not what is meant by a *complex* system. Take an organizational chart, for example. The flow of who reports to whom can seem confusing or complicated, but one way or another it can be mapped out. However, taking the same organizational chart and adding in the personalities, relationships, and history to the names on the chart reveals a complex situation. We like to think of it as: complicated situations can be solved or figured out; complex situations are navigated. The complex nature of a system implies layers of interconnectedness and interdependencies. Navigating through complexity requires a greater level of understanding, flexibility, and curiosity than the thinking required to sort out what is merely complicated.

21 JAN

Adaptive

We had just grasped a distinction between complicated and complex systems when another light bulb moment happened by adding "adaptive" to the concept. As Joss Colchester put it for *Complexity Explorer*, "Often used in 'adaptive systems,' the term 'adaptive' refers to interacting entities that individually or together are able to respond to environmental changes or changes between the interacting parts. The term can refer to a temporary modification to meet a changing context, or a long-term, permanent modification."

This was big for us. We had the epiphany that by understanding what complex adaptive systems meant we were understanding the foundation of "complexity speak." Yay us! All of a sudden, everywhere we looked we could see complex adaptive systems—systems that were always there, but it was like we saw them with a new lens. By adding "adaptive" to complex systems, we saw that systems are dynamic—that there is constant change, evolution, and movement within the system.

22

Influencing Pattern-Based Change

Denise Easton, co-founder and managing partner of Complexity Space Consulting, talked with us in Episode 117 Influencing Pattern-Based Change about how she was introduced to complexity science and how it influenced her life and work. She said, "A colleague of mine, Barb Siegle, always says, 'Complexity is something that is caught, not taught.'"

Easton described the story behind her book, *Complexity Works!: Influencing Pattern-Based Change in Teams and Organizations*, which she co-authored with Lawrence Solow. She said, "Having people recognize that surprise is information and is an opportunity and not a disaster is where we hope [the framework described in the book] will help people." Easton also said, "We thought one of the more interesting things about complexity science was how patterns influence and are influenced by the actions and interactions within the system." The framework she uses leverages a pattern-based approach to understanding and influencing organizational change.

23

Emergence and Higher Order

Emergence is what happens when things interact in new ways and produce unexpected results. However, there is more than emergence at play when it comes to complex systems. Cosmologist George Ellis explained in a talk given at the 16th Kraków Methodological Conference, "On the Nature of Causality in Complex Systems," that complex systems do a dance of give-and-take between higher level, top-down causation and bottom-up emergence.

We would all like for our world to make sense, so when it gets messy and unclear we often look for how we can restore order. A good place to look for change is at the surface—the day-to-day stuff, like how we humans interact with our environment. This is often where we can start to see disorder. And where there is disorder, there is emergence. However, Ellis explained that there's a ceiling to how much complexity you can get from bottom-up emergence before you once again encounter top-down causation. In other words, there's a limit to how much sway emergence (including innovation) has on a system before built-in mechanisms kick in and reinforce the higher structures upon the lower.

This happens in even the most innovative cultures and the flattest workplaces. Eventually, a higher level such as your founder's directive, your budget, prevailing attitudes, or regulations come to bear on what can happen next. And that's well and good because if no one ever drew boundaries anywhere to interact with what emerged, the whole works would fall prey to entropy.

Higher levels of order, which tend to be autonomous and more static, draw on the randomness of lower orders to create new possibilities, but they aren't controlled by them. For example, think of the organization's mission as higher order and the daily moods and inspiration levels of you and your colleagues as lower order.

We think eureka moments occur when, as if by magic, ideas from the lower and higher orders seem to arise mutually. This is where a complex system finds resilience, the dry land in an ocean of possibilities. In a dynamic system, resilience to small perturbations can lead to a kind of stability or coherence. The hierarchies of top-down causality and bottom-up emergence that make up a complex system are working together to produce possibilities that we can begin to identify. In other words, stuff starts to make sense.

In complexity-savvy organizations, we build top-down hierarchies to be flexible, allowing random innovations to help us revise the higher-level structures that steer the ship, but we don't become so enamored with emergence that we sail completely off course in search of sirens.

24

Emergence and Synergy

The New England Complex Systems Institute states, "Emergence refers to the existence or formation of collective behaviors—what parts of a system do together that they would not do alone." Or, as Joss Colchester said at Complexity Labs (now Systems Innovation), "Emergence describes a process whereby component parts interact to form synergies, these synergies then add value to the combined organization which gives rise to the emergence of a new macro-level of organization that is a product of the synergies between the parts and not simply the properties of the parts themselves."

At *Human Current*, we trusted and embraced emergence to guide our work. We continued to be in awe when the perfect guest presented at the perfect time. On several of our episodes we talked about how we trusted the process, which we believe allowed for us to recognize and appreciate emergence. We would even go as far as to say that *Human Current* was successful because of our philosophy and the value we placed on emergence.

25

Harmony

“Life is a game of hide and seek, because life is pulsation. On and off . . . all life is this flickering in and out, only there are enormous rhythms in it,” said Alan Watts in a *YouTube* recording, “Life is a Game of Hide and Seek.”

In this recording, Watts pointed out, “Lower orders are always being superseded by higher orders. The lower order can very rarely figure out what the higher order is that’s taking over, and may see it as a terrible threat—as total disaster, as the very end. But the principle of growth always has and always will continue.”

He described how what appears frightening and confusing may actually emerge as harmonious. “You may add up to some kind of sense that is not apparent to you in your ordinary consciousness,” Watts said. “What is, in other words, conflict at one level of magnification is harmony at a higher level. Life is a pattern of immense complexity. Each one of us is the delightfully complex undulation of the whole universe.”

26

Dance

In information theory, the smallest unit of information is called a "bit." It's easy to think in dualistic bits: yes or no, black or white, beginnings and ends. We think it's more useful to contemplate the dance of opposing forces, because the way they work together is the essence of change and creative tension. We create distinctions of meaning by knowing what a thing *is not* as often as by knowing what *it is*.

In a *TEDxUofM Talk*, "Mommy, Where Do Ideas Come From?," Steve Rush explored where ideas come from and ways to put opposites back together in new ways. Entropy breaks everything down to bits. In Newtonian physics everything is programmed to obey the same laws of degradation; to come down to the ground, for example, or to fall apart. But when one idea is strong enough, it gathers agency, becoming an attractor and pulling other bits to collect around it like a star. It gets gravity. It becomes a well, a world, a galaxy, a universe. It doesn't want to be alone. It's impossible to be alone, because matter and spirit are always dancing, notion and action pulling apart, coming back together, creating cultures, movements, change. The creating and the destroying are the same process. The oldest love story there is. The only game in town.

What matters is what's happening on the local scale, with you, with the systems you care about. The bottom-up activity. What is emerging as brand new from the dance of opposites doing their thing to reunify in your experience? When you get down to the very small, where information begins, things get weird. It can look like chaos (which all beginnings do). Beneath everything that happens, the very small are dancing and the bits of the world are considering how they will come together again. Seams are sewn with ideas. All of creation began with an idea, a word, logos.

You are a locus of activity, a swirling together of ideas. The rebellious dance of creation and destruction, of unruly quanta, is happening within you, and without, all the time, the beginnings and endings of a universe made of information. And whether or not you can make heads or tails of it when ends turn into beginnings or when disaster looms, change comes and growth comes with it.

27

Feedback Loops

Our complex environment provides constant feedback loops. As Herbert A. Simon said in *The Sciences of the Artificial*, "Human beings, viewed as behaving systems, are quite simple. The apparent complexity of our behavior over time is largely a reflection of the complexity of the environment in which we find ourselves."

Peter Senge's book, *The Fifth Discipline*, comes to mind when we think of systems thinking, personal mastery, and feedback loops. In our attempt to synthesize Senge's definition of feedback loops, we'd say that feedback is a reciprocating flow of influence, which is both cause and effect.

A significant insight from our experiences, along with Senge's book, is that feedback loops can be reinforcing and balancing, both negative and positive, and are influenced by patterns and behaviors. As he wrote, "The feedback perspective suggests that everyone shares responsibility for problems generated by a system." We especially recommend Senge's book if you're interested in learning about how feedback loops affect behaviors within a system, and how those behaviors can in turn impact the environment in which we find ourselves. As Simon said—quite simple!

28

Interdependency

As lifelong systems thinkers and growing complexity thinkers, seeing and understanding interdependencies came fairly easy for us. However, a big learning we had was in appreciating what interdependencies have to offer to a situation. We'll explain. We had this impression that seeing and understanding interdependencies would make navigating through a situation or relationship easier. Boy, were we wrong! You know that saying, ignorance is bliss? We think that in many cases, before we had an understanding of interdependencies, our ignorance made things easier to an extent. We could just "keep it real," do what we thought was best, and say what we wanted to say. With that mindset came all kinds of unintended consequences—which are great examples of feedback loops.

Because interdependencies exist within complex adaptive systems, there is no "easy" button. In fact, it's not about finding a solution; it's about navigating through a situation. While having an understanding of interdependencies doesn't provide a magic formula, it does provide a deeper level of understanding and appreciation to navigate through situations in a meaningful way. This requires a mind shift and, we believe, leads to a heightened level of mindfulness in which we appreciate the collective whole, not just the parts.

29

We can think of complexity science
and systems thinking
as really the sciences of connection
rather than the sciences of separation.

Jeremy Lent
Episode 074 The Systems Worldview of Human History

30

Describing Complexity as a Messy Human

We often forget that the very act of written and verbal communication tends to diminish ideas, especially complex ideas whose very existence is dependent on their context. Even our podcast format, which allowed time and space for deep conversations and storytelling, sometimes limited our ability to formulate and express ideas related to complexity. It is important to remember that our limitations make us vulnerable and human, and we have to recognize and acknowledge our limits in order to grow and make meaningful connections with each other.

Although much context is lost during the process of communication, connecting with our community and learning together is why we do what we do. We talked with and read literature by many different scientists, practitioners, and enthusiasts in the complexity science community and gathered diverse themes which encompassed, explained, and (dare we say) defined complexity science and complexity thinking. The following are some of what we heard:

Complexity science is:

- "The science of surprise," John Casti.
- "The science of open systems," Jean Boulton.
- Wholeness. "Complexity comes from the Latin 'plexes,' which means interwoven or entwined and basically something which is difficult to separate," Carlos Gershenson.

Complexity thinking is:

- Seeing and understanding how patterns change.
- Understanding how systems adapt, change, and evolve.
- Embracing paradoxes and the "messiness" of life.
- Integrative thinking, feeling, and sensing.
- Understanding how different levels of analysis interact with one another.
- Understanding system variation, innovation, and selection.

31

Communicating Complexity

Complexity thinking is not always intuitive. Linda Booth Sweeney said in Episode 116 An Educator's Guide to Systems Thinking, "We live in a world of systems, but for the most part we aren't taught to see them, to recognize them, to understand them, to make them visible or to work with them." She noted, "In school, many of us were taught subjects in a compartmentalized way," which limits our ability to practice systems thinking.

Yes, complexity can be challenging to talk about and define. It's like trying to describe energy. It is all around us and it affects our everyday lives, yet we struggle to articulate what it is because we weren't taught to see it. This means it is likely that there will be many misunderstandings and miscommunications, so we all need an abundance of tolerance from one another. Learning and personal growth are messy and ongoing processes, but you can find profound beauty in your journey if you practice openness, connection, creativity, and curiosity.

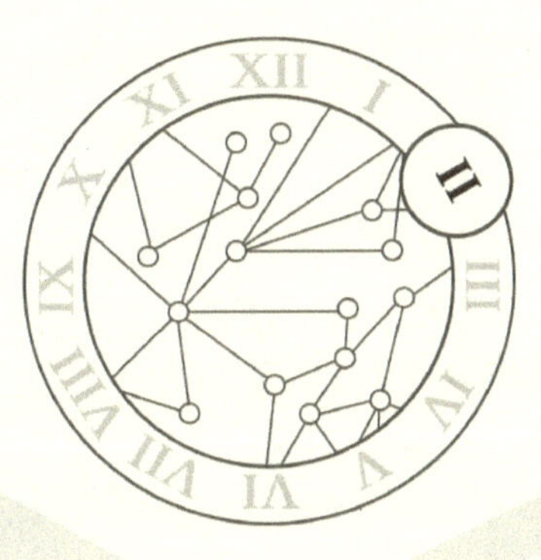

FEBRUARY

1

Opportunities for Greatness

Doug taught us that a healthy, high-performing organizational culture can be cultivated by providing people, communities, and organizations with the tools and opportunities needed to excel. A high-performing culture is a catalyst for greatness and changing the world. It's especially critical to cultivate high-performing culture in today's increasingly complex business environments.

When people, communities, and organizations are at their best, they can evolve into dynamically connected human ecosystems that allow the fullest expression of human capacity to flow through their every channel. As a result, these ecosystems can be conscious stewards of relationships and invested in outcomes, each other, and the greater good—ultimately changing the world.

Doug would tell us about the importance of understanding personal perceptions, group dynamics, and structural concerns in order to assess our cultural wellness. Since the stated and emergent cultures of an organization or community can sometimes differ, it's critical to get curious and ask questions to identify the gaps. To uncover even more valuable insights, Doug would tell us to identify the strengths, challenges, and opportunities for growth. We learned that this level of inquiry could impact an organization's or community's productivity, happiness, and alignment with its true purpose.

Doug taught us simple tools and methods for high performance. For example, he stressed the importance of point-to-point communication, which he said should be practiced as often as possible since so much of human communication is nonverbal. By meeting face-to-face, people can "read each other's energy fields," Doug would say. Face-to-face interactions also provide an opportunity to reduce incongruence and increase authenticity, which helps to foster trust. He stressed the importance of meeting regularly as a team and performing rituals to create rapport. He recommended that meetings should be about more than just solving problems. They should also provide reinforcement of shared purpose, a safe space to express opinions and perspectives, and a way to bring greater connection to a team. Additionally, Doug insisted on plenty of whiteboards around the office to foster creativity and collaboration, and checked that the whiteboard content would change often and not become stagnant. These are a few examples of how Doug influenced us to excel and create a healthy, high-performing culture so we could be a catalyst for greatness—and change the world.

2

History and Patterns

History plays an important role in our search for context and patterns. Patterns happen and change over time, so looking back and asking questions can help us recognize them and learn from how they evolve. As we see patterns repeat and build upon each other, we also begin to understand the context from which current social systems emerged, including many systems that now contribute to conflict and inequality in our society. Without a foundational awareness for the history of these systems, we will continue to unknowingly reinforce underlying structures that erode our social and economic equality. We also take our own power away when we refuse to acknowledge the part we play in reinforcing existing systems.

Peter Senge said in *The Fifth Discipline*, "Leverage often comes from new ways of thinking. In human systems, people often have potential leverage that they do not exercise because they focus only on their own decisions and ignore how their decisions affect others." One of the biggest lessons from complexity science comes from acknowledging that our behaviors are actually interactions which have ripple effects beyond our knowledge. Even as observers, we are not separate from the systems we observe. We will never fully comprehend complex dynamic systems—we can, however, begin to filter our thoughts through a lens of inescapable interconnection. Once we shift our consciousness to the complexity worldview, we will see reality for what it truly is and can begin to heal the broken pieces.

3

Looking to History

Humanity is in a crisis of consciousness. Society and the way we interact with each other is broken because we too often see the world in separate, static, and linear ways. Our reductive thinking keeps us stuck in the past, with no clear path for how to move forward, keeping us in constant conflict with the complex, dynamic, and evolving world around us. Luckily, an opportunity for change is presenting itself. Complexity science gives us a new approach to the art of inquiry, an art which can guide us forward. A complexity science approach allows us to explore and address questions beyond the scope of traditional science, including human behavior, cultures, and social movements. Complexity science pushes us to look at history, not for answers, but for a map toward a more positive future.

Every February in the United States we are presented with an opportune time to reflect on the past. Black History Month, which stems back to the 1920s as "Negro History Week," is here to remind us that the past contains important insights. Unfortunately, over the years we have missed too many of these insights and desperately need to expand our thinking in order to move forward in a positive way. With a collective shift in human consciousness, guided by missing lessons from our past and complexity science, we can work together to achieve positive change.

4

Seeing Systems

The scientific movement toward complexity is calling us to see the world in systems. Black History Month is the perfect time to embrace and challenge our new awareness. The potential richness of Black History Month lies in its call to action, asking us to reflect on a nation's history and ask ourselves what we have missed and what part we played in the system. These are common questions that systems scientists use to expand their thinking. We can choose to use complexity thinking and exploratory questions like these to open our minds, or we can continue to reduce the world to a series of separate issues to be solved in isolation, knowing this approach will keep us trapped in the past. Systems thinker and entrepreneur Tanuja Prasad said in Episode 118 Practical Applications for Interconnected, Nonlinear Systems, "It is the human desire for a simple, static answer to things that creates a contrast between what we want to see and what exists." With a complexity worldview we can use opportunities like Black History Month to expand our thinking, to see interconnections, to search for patterns and underlying structures, and to reflect on our own contributions.

Together, we can shift our focus away from static solutions that lead to unintended consequences and toward solutions that are emergent and co-created organically and collectively by the system itself. Reflecting on societal issues or historical conflicts with this new lens empowers everyone to create real, positive change in the future.

5

If you are really studying complexity theory and systems thinking, then you are being introduced into a deeply spiritual recognition that the world is interconnected and everything depends on everything else. . . . What spirituality means for me is to recognize that I am a minor, modest participant in a very large mystery.

Meg Wheatley
Episode 090 Reclaiming Leadership for the Human Spirit

6

Systems Theory, Racism, and Human Relationships

We interviewed T.K. Coleman, the education director for Praxis, to discuss how humans allow their conflicting mental models to influence the way they handle controversial topics, like racism. In Episode 025 System Theories, Racism, and Human Relationships, he quoted Jon Stewart: "We have made enormous progress in teaching everyone that racism is bad. Where we seem to have dropped the ball is in teaching people what racism actually is." Coleman also shared how understanding context and patterns within human systems ultimately empowers us to actively contribute to human progress. He said, "If human beings were only capable of treating problems that everyone was affected by, nothing productive would ever get done."

7

Discover Your Intercultural Intelligence

We interviewed motivational speaker, author, and leadership coach Gloria Burgess to discuss patterns and systems within human history for Episode 024 Discover Your Intercultural Intelligence. Burgess shared a Meg Wheatley quote: "When we are overwhelmed and confused, we reach for the old maps, the routine responses, what worked in the past. But to navigate life today, we definitely need new maps. The maps we need are in us, but not in only one of us. If we read the currents and signs together, we'll find our way through."

Burgess explained that because we are truly interconnected systems, how we relate to ourselves and each other matters. In this episode, we talked about a quote from Tim Wise, "When you allow racial disparity and institutional inequity to affect one part of the country, eventually it's coming back to get everyone." We all have the ability to gain intercultural intelligence in order to move through the world and systems differently. Burgess inspired us. She pointed out that the institution of racism was built with power and privilege—and people who built it can un-build it. She said, "Light comes in at the edges and end of a pattern, so a new pattern can begin."

8

Using Systems Thinking to Resolve Conflict

We had a conversation with conflict resolution expert Jason Dykstra about the complexity of conflict: why we're afraid of change, how to shift from a mindset of scarcity to abundance, and how understanding complexity in relationships can help us evolve ourselves and our world. Dykstra has used his knowledge of complex systems to help people and organizations find creative solutions to conflict and become "comfortably uncomfortable with approaching the unknown." He believes that personal growth affects the growth and evolution of an organization, and that no matter where you are in a system, if you make even small changes it will automatically cause others to change as well. Here is an excerpt from our conversation with Dykstra in Episode 003 Using Systems Thinking to Resolve Conflict:

***Human Current*:** How would you describe complexity in layman's terms?

Dyskstra: Basically, addressing the unknown. . . . For example, mapping out a system that a client is involved in might assist a person in understanding the various moving parts that their decision, action, or inaction contributes to the rest of the system. Or, simply asking questions to a person about how their actions affect the people around them, so that they can see the complexity of the situation and the ripple effects.

***Human Current*:** What benefits do you see from being a systems thinker?

Dysktra: I think a huge benefit is that you can step back and see the bigger picture. Being able to see all the moving parts and how they interact with one another allows you to have a better idea of the least intrusive way to assist a company and its leadership.

***Human Current*:** If someone wants to be more of a systems thinker or bring more complexity awareness into their life, what advice or actionable steps would you give them?

Dykstra: Step back and get curious. Often I think our initial reaction to anything is one of judgment. We judge what others have said, what folks have done, what our thoughts are, and what we are seeing. One thing that I have seen over and over again is that when we shift from judgment to curiosity, the entire world opens up. Also, often we get too close to a situation so that we can't see the forest through the trees.

9

Practice Conflict Resolution

In our conversation with Jason Dykstra about resolving conflict, he offered a simple three-step approach to taking a step back, creating space, and reflecting on our interactions with other people:

First, think of a recent moment where you encountered conflict with someone. Was your approach more reactive and judgmental? Responsive and curious? Reflective and reserved? Or something else entirely?

Second, think about all the ways you could have handled the situation differently.

Third, think about all the ways the other person(s) could have handled the situation differently.

When we think expansively and entertain multiple possibilities for behavior, we broaden our perspective. We realize how complex our experience is, and that within this complexity, there are abundant choices and paths to resolving conflict.

10

Part of what's most fascinating for me
is the opportunity in complexity.
The opportunity for growth, the opportunity for learning.
I mean complexity is all around us.
It's everywhere and it's every situation that we are in.
To some degree anything requires this adaptive
kind of framework. I think it's a great opportunity
for us to learn more about ourselves and
our role within this world.
Who are we and how do we contribute to this world?
It gives us this opportunity to dive deeper
into those things that we are interested in,
and to ask questions and build relationships.

Jason Dykstra
Episode 003 Using Systems Thinking to Resolve Conflict

11

The Power of Mindfulness as a Conflict-Resolution Tool

by T.K. Coleman for *Human Current*

It's easy to get angry or annoyed with people who don't share our points of view. It's even easier to get angry or annoyed when those same people insist on maintaining their points of view in spite of our best efforts to critique their opinions and change their minds.

Throughout the written record of human history, there has never been a shortage of frustrated thinkers who bemoaned the existence of irrational people. There has never been a shortage of proselytizers who felt deeply troubled by what seemed to be a unique epidemic of close-mindedness in their era. There has never been a shortage of aspiring revolutionaries who believed that the fate of the planet was headed for disaster because of all the so-called morons that refused to entertain new ideas. Our generation is no different in this regard.

We are a generation that longs to see change and progress, but like those who've come before us, we seem to face formidable resistance from those who seem wedded to the rules of conventional thinking.

"No one cares about the truth anymore."

"No one cares about real progress anymore."

"No one cares about basing their beliefs on facts and evidence anymore."

"No one cares about individual rights anymore."

"No one cares about the good of society as a whole anymore."

These are the kinds of stories we begin to tell ourselves after multiple failed attempts at getting people to open their minds to whatever it is we think they ought to believe. Such stories, although very tempting, are ultimately self-stultifying because they blind us to the most important truth about creating meaningful dialogue: You can never hope to change the world unless you adopt a charitable view of other people's ability and willingness to think intelligently.

The moment you make up your mind that people are incorrigibly stupid, your days of being an effective agent of social change are numbered. What does it even mean to invest in the future of our planet if that investment doesn't involve the people who inhabit the planet? And what does it even mean to invest in people if we don't take them seriously as entities that are capable of opening their hearts and minds to new vistas of possibility?

Changing the world requires a delicate balance between honesty and hope. On one end, we need to be honest about the ways in which our society is broken. On the other end, we need to be hopeful about our ability to create incentive structures that redirect human activities toward a constructive aim. Hope without honesty is delusional, but honesty without hope is paralyzing.

So how do we proceed? Do we sit back on our couches and try to change the world through positive affirmations? Should we all just purchase a copy of *The Secret*, collectively visualize a better world, and hope that the planet heals itself through the power of mental magic?

The answer to all the above is a resounding "no." The way forward is not delusional optimism, but mindful constructivism.

When many people hear the term "mindfulness," they think of things like Buddhism, meditation, being peaceful, taking walks, chewing your food slowly, etc. But mindfulness is more of a general orientation than a specific activity. To be mindful is to adopt a sense of life that says "In all things, I am student first and a teacher second." It is to maintain what Zen philosophers call "beginner's mind." It is to live with a recognition that creativity and problem-solving becomes infinitely more likely when we listen before we talk, when we look carefully before we judge, when we aim to be present with others before preaching at them.

If you want to make a positive impact on the world, I'd like to challenge you to be more mindful. Be mindful of the tone in which you express your ideas. Be mindful of your timing when you ask others to give attention to your ideas. Be mindful of the context within which you are sharing your ideas. Be mindful of the people you're talking to. Be mindful of their hopes, dreams, fears, and self-doubts. Be mindful of their needs and frailties. Be mindful of their unique experiences.

Be mindful of not just your capacity to teach them, but also their capacity to teach you. Be mindful of the fact that change takes time and that revolutions don't happen overnight. And above all, be mindful of the fact that you will always have infinitely more to learn than you will ever have to teach.

Having great ideas is awesome. Having phenomenal talent is awesome. Having strong work ethic is awesome. Having good character is awesome. Being mindful will keep all that awesomeness from going to waste.

Excerpted from a blog previously published at human-current.com/blog on February 10, 2016.

12

Humanity's Interconnection

When we really dig deep into understanding what it means to be human—a complex adaptive system with interacting parts and emerging collective behavior—we begin to see that our reality is interwoven and we participate daily in our own evolution, both individually and collectively. We cannot escape our interconnection. No matter how badly we want to believe we are separate, we are not. Humanity's greatest gift is that we are all in this together, all we have to do is accept it. As researcher and author Brené Brown says, "Connection is why we are here."

13

Human Networks and the Emerging Integral Society

Networks are powerful and fascinating, and more importantly, we believe understanding networks is crucial to social evolution. Why? Because understanding networks is fundamental to understanding what makes for healthy systems. We are living in the networked age, after all.

It becomes very evident how important networks are when you begin searching for a new job, for example. Many people are still trying to land their next career by searching through job postings, but a 2012 report from ABC News stated that 80 percent of jobs are landed through networking. This shows how important it is to invest in our personal networks.

In Reid Hoffman's *LinkedIn* article, "The Information Age to the Networked Age: Are You Network Literate?" he says, "I put myself at as many key intersections in my networks as I can. As a result, my network inevitably ends up connecting me with great entrepreneurs and great investments."

These examples prove that there is power within networks. Beyond investing in networks for personal gain, it is critical that we strive to understand them in order to expand collective consciousness and better the world. In other words, we are connected in more ways than we can currently comprehend, and there is immense power in these connections. This is the energy that binds us.

We can look to other living networks or systems, like photosynthesis, as examples for efficient and resilient ways we can cooperate. First, we must recognize that only once we shift our perspective and see the world as a web of complex networks and self-organizing systems will we transform and evolve away from the limiting, isolating, and competitive social structures that control our societies.

As Sally Goerner explained in her book, *After the Clockwork Universe*, "There are already visible signs of an emerging 'Integral Society' in which head, heart, and soul need no longer be at odds." Why not embrace it?

14

Meaning really is a function of the network. Meaning arises from the ways in which everything connects up.

Jeremy Lent
Episode 074 The Systems Worldview of Human History

15

Living from the Inside Out

We interviewed journalist and host of the *Happy Melly* podcast, Sam Mednick, for Episode 070 Living from the Inside Out. Mednick said, "You have to identify what makes you feel alive," and that "most of us haven't thought about our values." She talked about the importance of emotional intelligence and shared her philosophy on living a more happy and fulfilled life. Mednick quoted Tim Ferriss from *The 4-Hour Workweek*, "A person's success in life can usually be measured by the number of uncomfortable conversations he or she is willing to have."

16

I became interested in the complexity of things like social hierarchies, just to understand what structures, if any, support problems like social trust, as well as how they affect social resilience among different populations.

Josh Morgan
Episode 020 The Energy That Binds Us

17

Designing Meaningful Work in Complex Times

We interviewed Bud Caddell, complexity thinker and founder of NOBL Collective, for Episode 069 Designing Meaningful Work in Complex Times. Caddell explained how NOBL designs human-driven systems which enable teams to become more empowered, inspired, and engaged.

"I fundamentally believe that meaningful work is a human right," he said, and described how his understanding and application of complex systems science has influenced his mission to expand access to meaningful work. He spoke of "perpetual novelty," saying that when complex adaptive systems throw you constant novelty, "It's important not to use a playbook that was dated years ago."

18

Define Values to Live By

In order to successfully navigate complexity, we need a clear path with our feet grounded and our eyes fully open. The tricky part is we are the only ones who can create our path. No one can do this for us. Beyond mindfully showing up, we need to define our values, what is most meaningful to us, and live our lives with the intention of fulfilling our values.

Our intention is our clear path. As Robert Frost wrote, "The best way out is always through." The Barrett Values Centre states, "Your values reflect what is important to you. They are a shorthand way of describing your motivations. Together with your beliefs, they are the causal factors that drive your decision-making." Our values ground us with self-knowledge and provide a map for us to develop and grow in harmony with the world around us. It is also important to realize that our values might change throughout the different stages of our lives. We have to check in with our values as often as we can. Let's live from the inside out.

19

Search for Meaning

As time passes into history, individual narrative arcs will cascade, and the story that will be told is the one that made the most impact on the networks of human beings that digest it. You can't judge what will have meaning for someone else, but you can get better at following the trail of breadcrumbs that leads to meaning for you. So start anywhere. Start with whatever task, great or small, feeds your search for meaning. Adjust as needed. We know that in a complex system, changes in local conditions ripple outward to produce changes in the larger network. Your local experience is your best toolbox for speaking to the world.

20

We all have meaning-making skills
and we do that through narratives.

Sonja Blignaut
Episode 039 Company Culture as an Emergent and Evolving Phenomenon

21

Connect Through Storytelling

Humanity has been sharing stories since the beginning of time. Stories help us come together and make sense of the world. Complexity requires storytelling. Individual stories can inspire action and collective stories can shape cultures and values. Stories are the glue that hold our networks together. We all need to be listening, especially to voices that are underrepresented. We need stories from everyone, every walk of life, because without the whole story, we are all lost. We are lost without each other and our humanity is hidden among the missing pieces, the missing stories. If we want to glue ourselves back together we have to courageously listen without judgment and without motive. We also have to practice vulnerability and trust one another with our stories, realizing that our individual story is incomplete on its own.

In "It's an Interconnected World" for *Shambhala Sun*, Meg Wheatley wrote, "Everyone in a complex system has a slightly different interpretation. The more interpretations we gather, the easier it becomes to gain a sense of the whole." Let's share our story when others have earned the right to hear it and open our hearts and provide space for others to share theirs.

22

A Systems Worldview of Human History

We had a conversation with author Jeremy Lent for Episode 074 The Systems Worldview of Human History. Lent integrates systems biology, complexity science, and neuroscience with ancient wisdom traditions to explore how different cultures throughout human history have created meaning. He said, "We can use both traditional wisdom and modern complexity science as ways to try to get into some of the deepest questions about life that Western reductionist thinking seems to really miss through its structure of thought." Lent shared perspectives from his book, *The Patterning Instinct: A Cultural History of Humanity's Search for Meaning*, including how he believes humanity has gotten to where it is today and where it could be heading. He asks these fundamental questions about humanity's search for meaning, "Where do I get my values from? Where do I get my meaning in life?" He then stated, "We have to transform our values and the way we get meaning from the world."

23

Forget Permanence, Seek Resilience

Alan Watts said, "The more a thing tends to be permanent, the more it tends to be lifeless." Does permanence dictate that a thing be unchanging in essence? Matter persists through different states, and in its mutability and diversity it defies oblivion. Death destroys a star and creates a galaxy. Change is a force of nature that creates diversity through interrelationship. Without diversity, a system destroys itself and new forms fill its space.

Resilience is another matter—or at least, another state. Resilience is the ability to persist through adaptation. When something is resilient, it is not as easily disturbed by the fluctuations around it. When it faces competition, it must learn and become better. Its ability to change into something else grants it new life.

As human beings with a biological expiration date, we find resilience in networks through social bonds and the sharing of ideas because those containers perpetuate the contributions of our will and imagination beyond the scope of our biological lifespans. They live on beyond us, and in that sense, networks ensure our survival.

What happens when social bonds weaken and we entrust the agency of our networks to institutions? Can we say that these institutions reflect what we intend when we give them power? If not, what other ideas and structures (both internal and external) can we turn our attention to in order to build and evolve through diverse, networked conversations?

Understanding complexity, or at least
understanding how we can approach complexity,
is so useful in our everyday lives,
because every time we have a conversation,
every time we witness an action, for example,
we have different perspectives that come out of that.
And, that's part of complexity, right?
That we witness the same action, and
we walk away with two very different experiences.

Jason Dykstra
Episode 003 Using Systems Thinking to Resolve Conflict

25

Managing for Happiness with Complex Systems Science

Author, speaker, and entrepreneur Jurgen Appelo was named Top 100 Leadership Speaker 2018 and Top 50 Management Expert 2015 by *Inc.* and has become a key influencer in the workplace happiness movement. During our conversation for Episode 068 Managing for Happiness with Complex Systems Science, he said, "Sometimes it's just a matter of simplifying things in order to be able to address the complexity." Appelo discussed how practicing complexity thinking has influenced his career and management style. He also shared stories and advice on how to "manage the system not the people" with different tools, games, and practices. Appelo stated, "There is this realization in the world that we need to achieve more than just employee engagement. Employee engagement is important, it's about making people care about the work that they do, but we want a bit more. We want people to be happy. It's also about well-being and things that are not directly connected to the actual tasks."

26 FEB

Is Our Happiness at Risk? Reclaiming Our Humanity

Happiness is an emotion, a very humanizing emotion. There are many words people use to describe it, and sometimes it's almost indescribable. Defining happiness can seem as elusive as achieving it. Maybe if we simply look at what it means to be human, we can grasp the idea of happiness.

Being human—or humane—means benevolence, compassion, fellow feeling, kindness, consideration, understanding, sympathy, tolerance . . . basically, we are creatures driven by emotional connection. At the core of our humanity is our unique ability to love and have compassion. Whether we respond to it or not, we all have a constant emotional pull for togetherness and connectedness.

Think about all the experiences you have had in your lifetime. Chances are your most "human" experiences are the most memorable, because these experiences ultimately expand who we are and where we belong. They restore our inner peace, revitalize our freedom, and help us feel connected, so that we can become refocused on our inner being and relieved from the desire to change others (which is an impossible task and frustrating distraction).

Ultimately, it is our humanity that brings us joy and happiness. It gives us a fulfilling purpose that we can achieve through simple actions, actions which highlight the best version of ourselves. If there truly is a connection between what it means to be human and happiness, we need to ask ourselves: What are the consequences that might arise from neglecting our humanity for progress and efficiency? Is life in the fast-lane and the quest for machine-like efficiency putting our happiness at risk? Gianpiero Petriglieri answers this way in "How To Make Technology More Human" for *GPDFBlog*: "Once all we care about is efficiency, and humanism is reduced to a matter of style, the real threat comes from the smart machines that we have become, not from those we will build."

27

The Energy That Binds Us

In Episode 020 The Energy That Binds Us, we discussed the meaning of our name, *Human Current*. It encompasses the complex connections that shape and influence us: our evolving human mind, our personal and professional networks, our ideas, and our history. We each play a role in the development and evolution of our humanity and the energy that binds. We introduced sociologist and fellow podcaster Josh Morgan, who hosts *The Plural of You* podcast. He told us, "I try to encourage everyone to see humanity as this whole, but at the same time recognize that we are all constituent parts of this whole. We all have a role to play whether we like it or not."

28

It's Not All About You

Winning and success are admirable goals in most of Western culture. Maybe a little too admirable. Competition is necessary and it is just as important as cooperation in complex systems, but problems can arise when winning becomes *more* important than cooperation. The context in which competition exists is important. As our guest Laurie Marshall said in Episode 051 Learning from Nature, "Competition in nature is always in the context of an overriding, interconnected mutualism where all organisms benefit from keeping a balance." Balance is key.

Human ego can get in the way of our cooperation and can stifle system goals because our ego wants to win, above all else. Those who overexert their ego and focus on winning are playing a finite game because once they win, the game is over. On the other hand, those who focus on cooperation and the system's reason for being are playing the infinite game, because they play to keep the game going.

At the WorkHuman 2018 conference Simon Sinek said, "Business is an infinite game," because there's no such thing as "winning" in business. There is also no such thing as winning in life. We are all a contributing member of an ecosystem that is much larger than us. We contribute to this system, we do not control this system.

There is no ego in eco, as much as we want to believe otherwise. Alan Watts said, "By replacing fear of the unknown with curiosity we open ourselves up to an infinite stream of possibility. We can let fear rule our lives or we can become childlike with curiosity, pushing our boundaries, leaping out of our comfort zones, and accepting what life puts before us."

So, let's embrace uncertainty, play the infinite game, and let go of our egos, because life will happen in spite of us—not because of us.

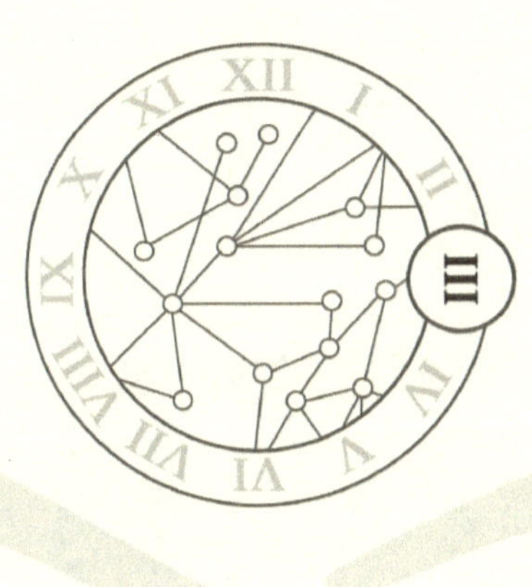

MARCH

1

Owning Your Strengths

Doug often talked about alignment of teams and individuals. He wholeheartedly believed that when you play off of your natural strengths and genius zone, you and your team will have far better results. Doug would share an illustration of a centipede and point out all of the centipede's legs, saying, "Each leg has independent actuators and independent feedback loops, and they are all part of the same system of goal setting and monitoring." Using the centipede metaphor, Doug taught us that when the skills of individuals and teams are aligned they move almost effortlessly "from a rough surface to another rough surface with very high assurance."

Because Doug viewed everyone as having agency within a larger system, he valued the people he hired as empowered agents in his company—like legs on a centipede. To ensure that we would align and work effectively together, Doug had us complete the Enneagram of Personality Types. Discovering our results was eye-opening and increased self-awareness, and learning about teammates' types helped us be more understanding of each other. Doug believed that by having this awareness, like the legs on a centipede, we would be able to work more effectively together and leverage each other's strengths. We learned that while collectively having the various types of personalities on a team can be challenging at times, overall it is rewarding and ultimately strengthens the team as a whole.

Doug's ideas around alignment reminded us of a concept presented by Malcolm Gladwell in his book, *The Tipping Point*. Gladwell described three archetypes of people defined along a spectrum. These archetypes are the connectors, the mavens, and the salesmen. Connectors are people specialists. They know and stay in touch with many different people and bring the world together by connecting people. Mavens are information specialists. They want to educate, not sell. Salesmen are charismatic. They build instant rapport and make others feel good with their lively spirit.

By knowing our own strengths and fostering each other's, we can work together in more powerful ways. This alignment helps teams leverage their ability to change the world.

2

Culture and Human Systems at Work

Let's eavesdrop on part of Episode 005 Culture and Human Systems at Work, while co-hosts Stacy Hale and Angie Cross discuss the workplace as a complex living system:

Hale: Rousseau argued back in the 1700s that nature is not stagnant, but alive and dynamic. People like to think that A causes B and B causes C, but that's another way of saying reductionism. Reductionism fails in living systems because there are so many moving parts, especially when people are involved, and even more so when an organization is involved. In organizations, how do leaders balance top-down control and bottom-up emergent behavior? This is something we talk about a lot in complexity theory. It's possible that if you wanted to understand the culture and behavior of your organization, you'd look not just to the leaders, but to the relationships.

Cross: Yeah, not just the relationships in the organization, but the relationships that we naturally bring into the organization. For example, I'm a mom, wife, and daughter. I don't just come into work as an employee. I am accountable to these relationships outside of the workplace. I can't always turn off my "mom hat" in the middle of the day. We bring all of our relationships into the work environment and they are also an extension of that living system.

Hale: How do leaders not get confounded by the fact that there is no way to completely control the culture of their organization? It's in many cases so emergent. How do you not pull your hair out at the end of the day when you realize all the moving parts that are potentially affecting the climate of your organization, its productivity, and the attitudes of its employees?

Cross: You just used a phrase that I had a physical reaction to: "control the culture." You can't *control* a culture, you can *foster* a culture.

Hale: There's a complexity theorist at the Santa Fe Institute, Geoffrey West, who said all human interaction is just one social network. It made me think of that when you said you can't control a culture, but you can foster it.

Cross: Having a systems thinking lens allows us to think outside of ourselves and see how someone else's interactions affect us. Having a systems thinking lens allows us to see what we're capable of in redirecting and shifting culture, shifting energy, and growing our empathy at the workplace as well. Whether we know it or not, we're all connected.

Hale: I just keep coming back to feedback loops, and the way we interact with all systems is a feedback loop. And so change on a global level is based on this interdependency, the individual parts interacting. You could change a whole organization based on feedback and so if it's fluid, if it's clear, and the lines are clear and feedback is incorporated and not interrupted by leadership wanting to control the organization, then what would it do? Assuming that people who work there are aligned with the organization, that they share the same values as were set forth with the organization's inception, what happens when everybody moves as one?

Cross: What if everyone is aligned, but what if they're not? What if you could be the agent in that organization and want to redirect it?

Hale: OK. So, let's say that's true. How is that not controlling things? Where's the balance there?

Cross: I'm a big believer that the only one we have control over is ourselves. I can only be responsible for my own character and at the end of the day, who do I want to be?

3

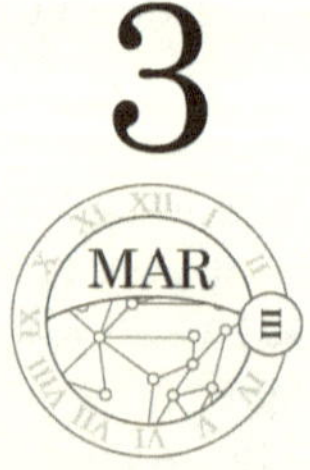

Leaders Help Shape Culture

The tricky part for leaders is that a culture is made up of numerous systems working together. Leaders have to confront all the underlying systems which shape culture including values, support systems, processes, feedback, and reward systems. It's daunting, but not impossible. Recognizing that all these systems influence and support company culture is the first step to change. It is also important to recognize that culture is not the "job" of one department, or a couple of leaders. It is a collective effort, co-evolving with people and supported by heuristics or guiding principles. As Orrin Woodward said in "Systems Thinking and Culture" for *Toolbox for HR*, "Just as scientists, doctors, veterinarians, and engineers study and create systems, leaders must study, change, and create cultures. . . . Think of culture in an organization as a riverbed, directing the flow of the river. The riverbed sets the boundaries of acceptable behaviors, controlling the flow of the water, helping new people move within the cultural river."

4

I believe that culture emerges. It's not a thing, but an emergent property of a complex adaptive system.

Sonja Blignaut

Episode 039 Company Culture as an Emergent and Evolving Phenomenon

5

Empowering Individuals

Everyone within a culture needs to feel valued and empowered enough to act and influence others. Starting with the individual is important to strengthening any type of company culture, and it is especially true for a wellness culture. We all have a role and responsibility for our own health. It's a very personal matter.

As reported by the Society for Human Resource Management, the number one reason employees don't participate in company wellness programs is because they think they can make changes on their own. Leaders can reinforce positive beliefs by empowering employees with knowledge, access, and support, rather than ignoring good intentions and trying to coerce participation. Examples of empowering actions include access to healthy food, a company gym, and a company-wide healthy-living newsletter. Support can also be as simple as setting good examples and providing encouragement.

6

Align Reward Systems with Desired Behaviors

Microcultures, rituals, and traditions are systems which exist within a culture and strongly influence behaviors, including those that affect our health and wellness. Since people tend to conform to norms in order to satisfy their sense of belonging, leaders have a responsibility to influence or shape social norms by rewarding desired behaviors and discouraging undesirable ones. Leaders who set behavioral boundaries ultimately direct the flow of the cultural behavior. According to Orrin Woodward in "Systems Thinking and Culture" for *Toolbox for HR*, it is important to keep in mind that "unless the reward system lines up with the culture, it will not change behavior."

The behaviors we are regularly exposed to will shape our own, so even wellness can be contagious. Scientists have observed the phenomenon of copying behavior, from runaway positive feedback loops in economies, to preferential attachment in social media, to homophily in public policy. The question to keep in mind is, how can we make the most of this tendency within complex systems to encourage beneficial patterns, such as healthy cultures and lifestyles?

7

Probably the most crucial concept to convey to people is the notion of emergence and in particular the idea that a complex system is one that consists of many different elements. The idea is that each individual has a simple behavior, but by interacting with each other and with their environment, the behavior of the system as a whole emerges from all of these interactions. And, that behavior can sometimes be very unpredictable even if you know what all the individuals are doing.

Paolo Gaudiano

Episode 048 The Intersection of Diversity, Innovation, and Performance

8

Identify Leverage Points

In "Making Systems Thinking More Than a Slogan," published at *Network for Business Sustainability*, John Sterman said, "Our guiding values offer the most important leverage point for sustainable change, and a small change can lead to a large shift in behavior." Speaking not only to the company's guiding values, but also to individual value systems is a powerful tool for impacting human behavior. Leaders can create a framework for company values, while also finding space to accept and respect the values of everyone.

A values exercise is a great way to capture this information. Here's one interactive approach to values clarification:

1. Ask employees to write their own top values on sticky notes.
2. Then ask them to share their values as they place the sticky notes on a board or empty wall.
3. Once everyone has shared their values and all of the sticky notes are visible, discuss how they are nodes in a system.
4. Using string and pushpins or drawing lines on a whiteboard, connect the nodes to find the connections in the network.
5. Wrap up this activity with the following debrief questions:
    - How was this activity for you? Talk about your ah-ha moments.
    - How can we use what we've learned from this activity?
6. Finish by thanking everyone for sharing their values and helping to map the connections between nodes.

Connecting individual values to company values gives leaders the ability to think about and create goals which will shift behavior in a positive way.

9

Inspiration and Emergence

Douglas in his office, beside pictures of archangels.

We wandered down the hall to Doug's office where we first started recording our podcast episodes. We loved spending time in Doug's office and taking in all of the books and the pictures of archangels that hang on his wall. On a quest for inspiration we stumbled upon an old beat-up copy of Erich Fromm's *The Sane Society*, which was wedged between a book on dynamic control systems and *The Essential Jung* by Carl Jung.

For some reason *The Sane Society* stood apart, and we didn't know why until we opened it up to a random page. On page 177, Fromm questioned "the meaning of work" in our capitalist society: "But work is not only an inescapable necessity for man. Work is also his liberator . . . his creator as a social and independent being."

Fromm was suggesting that we work to create meaning, and in that active search for meaning, the work also creates us. That's emergence! That's when work feels like magic.

10

In complexity, context is everything
and there are no recipes.
You really need to understand
what the emergent contextual ideas are,
and not just come in and apply your best-practice recipes.

Sonja Blignaut
Episode 039 Company Culture as an Emergent and Evolving Phenomenon

11

A Philosophical Look at Company Culture

Our guest for Episode 038 A Philosophical Look at Company Culture was Benjamin Taylor, a systems thinker and business evolutionary. He shared a metaphor about the concept of control, attributed to systems engineer Paul Plsek. The metaphor contrasts the act of throwing a stone with releasing a live bird. When throwing a stone, the thrower can generally control the direction and distance of the thrown object, so results can be projected and achieved without much uncertainty.

On the other hand, when releasing a live bird, the thrower has no control over the bird after it leaves his hand. There is much more ambiguity about where the bird will end up. Plsek's metaphor suggests that instead of trying to control the bird, the person releasing it could place a bird feeder in the direction he or she would like the bird to fly. Or, maybe if the bird is released in an environment with boundaries it could be successfully influenced and directed. Likewise, although it is nearly impossible to control a complex adaptive system made up of humans, there are still ways to influence it.

For leaders within an organization, this influential power is a big responsibility. Lisa Cash Hanson, chief executive officer of Snuggwugg, defines leadership as "the ability to guide others without force into a direction or decision that leaves them still feeling empowered and accomplished." Leaders are responsible for enabling people while also setting boundaries; and even more importantly, they are responsible for themselves and how they show up in the workplace.

12

Applying Complex Systems Science with WorkHuman

For Episode 076 The WorkHuman Movement, we interviewed Mike Wood, social marketing manager at Globoforce, a company dedicated to creating more human-centered workplaces. Wood shared how Globoforce is pioneering the WorkHuman movement "to galvanize leaders worldwide to harness the transformative power of people for the next generation of human resources."

Following our interview with Wood, we attended the WorkHuman 2018 conference in Austin, Texas. Their speaker lineup was a major draw, for it included well-known people who have exhibited leadership in their fields: Brené Brown, Shawn Achor, Salma Hayek, Simon Sinek, and Amal Clooney, among others. As sparkling as this lineup was, there were even more compelling reasons for our excitement. We had a desire to learn more about the WorkHuman movement, be part of their mission, and contribute to the conversation with our own testimony about how complex systems science can help transform workplaces for the greater good.

We looked for patterns or themes that underpinned the event and the messages that were presented. We realized there are many underlying factors that guide our humanity, personally and professionally, and ultimately there is no divide between our home and work lives. What we need more of in the workplace is what we need more of in life. We need more real connection, tolerance, trust, equality, diversity, resilience, respect, flexibility, vulnerability, healthy boundaries, and mindfulness. All of this adds up to greater wholeness, wherever we are.

13

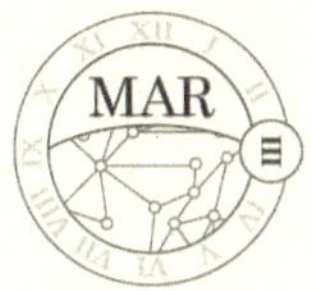

Always assume a posture of curiosity or not knowing. Be curious. Question assumptions.

Sonja Blignaut
Episode 039 Company Culture as an Emergent and Evolving Phenomenon

14

Culture and Human Systems at Work

As our podcast was still taking shape, we began imagining the workplace as a complex living system, musing on how information travels through networks and getting meta about culture and change. The following is an excerpt from Episode 005 Culture and Human Systems at Work, with co-hosts Angie Cross and Stacy Hale.

Hale: We're learning that there's greater complexity at work than what is implied through a position of authority, or no authority, or a managerial dictate. Think about it. . . . I give some person encouraging words as I'm walking by after getting my coffee. If you acknowledge the complexity of a human system you don't know how far that's going to carry. That person could just happen to be going in to lead a meeting, and it could make or break their interactions for the day and they could be a key to someone else's experience. We just don't know, but complexity is not our enemy, it's not something to be, in my opinion, slain like a dragon and made into simplicity. I think that's dualistic thinking, I think that's either-or thinking.

Complexity is a fact of life, it is literally a fact of biological life. And so no wonder it's at work in our organizations and we can make the most of it.

Cross: I love that, and that you can feed into someone else's day with positivity.

Hale: Yes, nurturing those feedback loops when they're helpful. Be intentional about them.

Cross: Yeah, because we don't just exist in silos, we exist in this system. . . . Some people hear "complexity" and they want to run away from it. We even had a conversation around how complexity doesn't sell, who would buy complexity. But it's really about embracing it, understanding it, and then you get to see the layers in it, and the richness, and noticing it for all of its facets. We don't need everything to be simple.

Hale: No, we could play our parts beautifully. Think about a symphony, a lot of moving parts. If you are a bassoonist you don't have to know what the French horns are doing over there. Complexity is inevitable and I think it's what gives beauty to our greatest natural systems. By acknowledging it, by understanding complexity theory, we give credence to the idea that small actions can create big change. Think about who all you could be affecting and not even know it.

15

Relationships and Networks

A company's brand is an ecosystem co-evolving with stakeholders, employees, and customers. It goes way beyond a product or service. As co-host Stacy Hale said in Episode 017 Emergent Ecosystems, "A company is really defined by its people and the current of information that runs between them." Neglecting to pay attention to relationships and networks within a brand system (or any system) can lead to disastrous results.

In today's world people have many choices, access to more information, and more ways to buy. This constantly changing, intensely competitive landscape demands that companies value thought leadership, authenticity, and relationship building. It's all about trust. Why should people trust you?

Many companies struggle to maintain control over their reputation because they treat problems in isolation without regard for their networks or the valuable feedback their networks have to offer. This reductionist approach is too rigid and can restrict a brand from reaching new audiences or adapting to new markets. By staying grounded in their value structures and trusted networks, companies have the ability to embrace change and ambiguity.

Brands can rely on their identity to form and strengthen relationships, and by focusing on relationship building, they seem more relatable, more human. Customers relate to what companies think, stand for, and represent. Without consciously realizing it, people actually feel connected to or compelled by the underlying value system that defines a company's brand.

Top brands take a systems approach to creating their identity and realize that every point of interaction matters. The goal is to "practice what you preach" through seamless experiences, so that every person who interacts with the brand, both internally and externally, receives the same level of quality. As Raphael Bemporad said in "Creating Brands of Enduring Value" for *SustainableBrands.com*, organizations will win "by creating brands that prize authenticity, build deeper relationships, empower seamless experiences, connect us to trusted networks, and treat us all—men and women—like humans."

16

Allowing for Emergence at Work

Our team at *Human Current* was purposefully flat and self-organized. We incorporated each team member's individual search for meaning, knowing that what they found on their own "treasure hunt" would add far more to the collective richness of the conversation. Here are some ways we allowed for emergence at work:

Look for serendipity. Emergence means being open to that which comes to us from outside of ourselves. Doug reinforced our tendency to look for what is spectacular in everyday life, whether from a page in an open book or a chance encounter with a person. Doug gave us advice to understand the totality of knowledge in the universe as a "source field" from which infinite connections and human progress can be made. Today's science speaks of a fundamental interconnectedness down to the quantum level, so we don't think Doug's claim was too far off from today's reality.

Spark up a conversation between you and your network. Feel inspired by your local network and with the greater network of information and relationships beneath all things. We think inspiration speaks to you every which way it can. An idea whose time has come emerges not in a flash of lightning, but at the tip of a network of exchanges and unlikely connections—conscious and unconscious—and most often in a team setting. We are all holding a piece of the puzzle, and the only way we can get to the big picture is to see how they fit together. Humans weren't meant to work in a vacuum. We're meant to collaborate and co-evolve together.

Have a beginner's mindset. In creative work, a beginner's mind will always have the advantage, so do whatever you must to get there. The best work you can do, in our opinion, is to develop good listening skills, seek out and destroy your unconscious biases, and realize that novelty is produced by mining the relationship of seemingly unrelated nodes. Go ahead, hang out with strangers, talk about everything with people you love *and* the ones who drive you nuts, let the tension of your burning questions eat at you. Have random conversations. Let it simmer in the background as you go about your day. Then pick up a book, any book, and turn to any page. See what happens.

17

Responsible System Leadership

In Episode 038 A Philosophical Look at Company Culture, we asked Benjamin Taylor how much control does a leader truly have when it comes to shaping company culture? He responded by describing company culture as "the lived experience of the people" that leaders can't control. In fact, it isn't a *thing* at all; rather, it is a set of intended and unintended consequences which result, in part, from leadership behavior. He emphasized that when leaders seek to shape or engineer a company's culture, they are seeking an unattainable level of control.

If leaders have little *control* in shaping company culture, we wondered if they have the power to influence it at all. The term "influence" means the capacity to have an effect on the character, development, or behavior of someone or something. Taylor said, yes, leaders have significant influential power on the people within an organization, which ultimately affects the company's culture.

Leaders only have control over their own behavior, but people look to them for guidance, social cues, and support. Therefore, leaders have to remain mindful of the responses that result from their actions and decisions. Rather than seek an unattainable level of control, Taylor urges leaders to recognize and accept the intended and unintended consequences that might arise from their actions.

18

Cultures, by and large, are what they are
because of the interactions and relationships
and stories carried in these cultures.
The final product is an emergent property.
It's a whole that is more than the sum of its parts.
It's dynamic and not predictable,
therefore it's not controllable and undesignable,
but you can tune in to what patterns
create conditions conducive to life,
and you can try to make design interventions
and set the right intentions behind your design
to aim to design for positive emergence.

Daniel Wahl
Episode 122 Design-Led System Transformation

19

Company Culture

Sonja Blignaut was our guest for Episode 039 Company Culture as an Emergent and Evolving Phenomenon. Blignaut is the founder of a niche consultancy called More Beyond. She shared how using a complexity thinking approach to study and influence company culture helps leaders avoid best practice "recipes" and reach positive organizational change. She said, "Hold your plans and outcomes very lightly. Focus on having a clear direction and a purpose but not an alignment to a specific goal."

Blignaut shared two quotes she especially relies on. The first is, "Think anew and act anew," by Abraham Lincoln. The second is Mark Twain's words, "It ain't what you don't know that gets you into trouble. It's what you know for sure that just ain't so." She discussed her realization that "in complexity, diversity really matters," so she surrounds herself with people "who have radically different perspectives" than she does. Blignaut also stated, "The moment we think we know the answer, or even the question, we close down our options to explore and we can no longer find the things that we didn't know to look for."

20

What Gets in the Way of Our Authentic Selves

Over the years, we had several conversations about what people need in the workplace and in life. We learned that a profound lack of systemic thinking, a desire to control, and a debilitating fear of uncertainty all reside at the core of what is broken in our humanity. We are not facing reality together. In the saying quoted by Anaïs Nin and others, "We don't see things as *they* are, we see them as *we* are."

Through our conversations and interviews, we explored what gets in the way of us being our best selves, and we discovered seven themes that can limit us: command-and-control leadership, fear of uncertainty, reductionism, linear thinking, separation and disconnection, scarcity mindset, and rigidity.

To truly grow and be our best selves as individuals and as a society, we need to be conscious of what's getting in the way. In any given situation, we can ask ourselves if any of the limitations listed above have emerged and, if so, how and what can be done to eliminate or reduce the limitations.

21

If you want to create a human-centric workplace,
there are a lot of ways to go about it,
but there is one thing that every company must do,
and that is to be real and authentic.

Erica Keswin
Episode 080 Bring Your Human to Work

22

Work Whole, Work Human

Many of the speakers at the WorkHuman 2018 conference shared their wisdom on workplace appreciation, human connection, and empowerment. There was an overarching emphasis on shifting work culture toward bringing your whole self to work.

Bringing our whole selves to work is essential for us to thrive as humans, and for companies to survive in today's environment. We were thrilled (and relieved) to learn that there are thousands, perhaps even millions, of people who also believe that our whole selves cannot be separated into parts, and that this holistic approach is increasingly being practiced in companies around the world.

At the conference, speaker Shawn Achor said, "Once we realize that our energy is interconnected, we see that the more you create across your ecosystem, the more potential you will unlock." Just as our interconnectedness and interdependence makes up our global society, all of our own experiences and interests make up our own unique, complex tapestry.

We agreed with speaker Sheryl Sandberg: "Bring your whole self to work. I don't believe we have a professional self Monday through Friday and a real self the rest of the time. It is all professional and it is all personal."

23

Relationship Building and System Innovation

We had a conversation with entrepreneur, public speaker, and systems thinker Amber Christian for Episode 083 Relationship Building and System Innovation. Christian discussed how to build strong business relationships through personalization, technology, journey mapping, and transparent processes. She stated, "The end goal should be to work ourselves out of a job." She also discussed using systems thinking to understand whole processes, identify unintended consequences, and strengthen a network of connections. "If complexity thinking is done well, it leads us to a clearer understanding of the choices we are making as leaders."

24

Social Evolution, Trust, and Happiness

How often have various philosophers and religions told us about the unseen connections between humans and the collective identity of humanity? We should know by now that there is a profound energy that runs between all of us. It is a complex, social phenomenon.

There are many complex connections that shape and influence the human current that runs between us: our evolving human mind, our personal and professional networks, our ideas, and our history. We each play a role in the development and evolution of our humanity and the energy that binds us.

Our lives have always depended on other humans, and our dependencies on one another have grown with time. As the world population increases, people adapt by expanding their networks, establishing more relationships, and joining more social groups. According to Pascal Vrticka in "Evolution of the 'Social Brain' in Humans" for *Huffington Post*, "The increasing size of such expanded social groups in turn entailed further social brain development, ultimately giving rise to the present form of the human social brain."

Our ever-changing world demands that humanity keep up with social evolution; therefore, we are now equipped with a larger neocortex in order to carry out and maintain more complex human interactions. This social brain evolution has gradually altered our needs and contributed to how we search for happiness.

The World Happiness Report of 2012 examined personal and national variations in happiness. The research recognized that, no matter a person's nationality or lifestyle factors, "We are social animals through and through. We learn through imitation, and gain our happiness through meeting social norms and having a sense of belonging to the community."

Over time, human experience has taught us that mutual social interactions are rewarding, while social rejection or disapproval can be painful, even causing physical pain. Therefore, social trust has become a critical factor in our social resilience and happiness. We must ask ourselves: How are we identifying with and showing up in social systems? Are we nurturing our social brains and projecting trustworthiness to others? As Josh Morgan said in Episode 021 Cultivating Social Resilience, "The most realistic approach to building trust is to be trustworthy, and to make conscious efforts to help others where possible."

25

Bring Your Human to Work

We interviewed Erica Keswin for Episode 080 Bring Your Human to Work. Keswin is the founder of the Spaghetti Project, which was inspired by a 2015 Cornell study showing that firemen who eat together are better at their jobs (they save more lives!). Keswin stated, "The Spaghetti Project brings intention to why connection is important for us as people, for our businesses, and for the world." It was clear to us that she is passionate about the science and stories of connection at work. She said, "I have always been a connector. It was something that I felt and knew intuitively, but now there was science to start to back that up." She also revealed that, "Left to our own devices, we are not connecting."

26

Happiness Is Only a Walk Away

Have you ever wondered what happiness literally looks like in your brain? The "Happy Cell" video found on *YouTube* is of "a myosin protein dragging an endorphin along a filament to the inner part of the brain's parietal cortex which creates happiness." This cellular process literally looks like taking a walk. Is it ironic that cells in your brain take a walk in order to create happiness? We don't think so.

Taking a walk gives you a chance to take time out for yourself to think and reflect. It allows you to connect with your body promoting deeper breathing for relaxation and stress-reduction.

It also triggers a number of other processes which can contribute to your sense of happiness and well-being, including:

- Promoting the release of endorphins, the "happy" hormones.
- Releasing adrenaline, the body's own mood lifting chemical.
- Producing hormones to improve sleep.
- Releasing muscle tension.

Walking really can make you happier. Even the cells in your brain know this! Getting out of your head and into your body is a form of mindfulness. Energizing your body allows your brain to focus. It's unfortunate that often we neglect the body in favor of the mind, without recognizing their interdependencies. We can easily remain stuck in our heads. When we do, we are not fully experiencing life, and life is where ideas live.

Active mindfulness is a powerful way to spark creativity, and by active mindfulness we mean being physically active and mentally tuned-in. This way your body and your mind are energized and your awareness level is at its peak. Now you are primed to reap the ultimate benefit from your mindfulness practice: guidance from your subconscious. Creativity thrives in the subconscious.

A research paper published in the *Journal of Experimental Psychology: Learning, Memory, and Cognition*, "Give Your Ideas Some Legs: The Positive Effect of Walking on Creative Thinking," by Marily Oppezzo and Daniel L. Schwartz, stated, "Walking opens up the free flow of ideas, and it is a simple and robust solution to the goal of increasing creativity." Creativity and happiness are inarguably linked to each other, and the promising truth is that they are accessible to anyone who is willing to take a walk.

27

Organizational culture is a catalyzer of performance. If negative, culture can be an inhibitor.

Oleg Konovalov
Episode 037 The Complexity of Living Organizations

28

Trust the Process

Understanding complexity shouldn't make people anxious; it should, paradoxically, make life simpler. With all this talk of emergence and working with uncertainty, there's an element of trusting the process at work. What does that look like? In Episode 008 Trust the Process, co-hosts Angie Cross and Stacy Hale discussed how to structure your attitude, mental models, and workplace for emergence. Hale said, "By being aware of our mental models, we can see how everything that happens is filtered through us, and we can guide that."

We wove big data into our conversation and Cross posed the question, "How do you use big data and yet still take into account heuristics and passing on knowledge?" Hale acknowledged, "The wisdom that your grandmother shared with you, on the porch with a glass of iced tea, is still important." She added, "Even though we're becoming complexity-aware, and we're understanding the importance of networks in the evolution of our societies, economies, and selves, we're still clinging to our big data to draw reductionist conclusions."

We explored how understanding life as networks and an awareness of our own value systems and mental models bring us a sense of personal mastery. Hale said, "Maybe it's not so much trying to achieve the ultimate certainty that we need to be concerned with, but being responsible with certainty, knowing what our limits are to what we can know, and then leaving the rest to practice."

29

Art, Complexity, and Human Evolution

In Episode 044 Making Sense of Human Systems, we talked with Dave Snowden. He said, "Art comes before language in human evolution." This got us thinking about the role that art plays in our lives as humans, why it's important, and how it influences us.

In our search for answers we came across a conceptual artist who has influenced American culture with his works of space art. Robert McCall was an illustrator for *Life* magazine and according McCall Studios, "He has done more than any other artist to enable Americans to visualize their nation's presence in space. He is described as an optimist and dreamer, unrestrained by scale, unlimited by location, unconfined by the boundaries of what is known or what can be imagined." McCall was so much more than an artist, he was an innovator and leader because he inspired others by communicating his ideas with hope, imagination, and enthusiasm.

Through McCall's art, we discovered that art can be a profound way to communicate brilliant ideas. Visionary art can translate complex or "zany" (as Doug would say) ideas into something tangible and relatable by tapping into the imagination. We believe art is one of the most powerful innovation tools. Its power is derived from its limitlessness. There are no rules or boundaries and no idea is too far-fetched for art. Therefore, art is free to nudge at the boundaries of society with such raw honesty that it propels and expands human consciousness.

Human progress demands that we embrace the limitlessness creative freedom art offers. Leaders can embrace creative freedom in the workplace by replacing fixed rules with guiding principles so that innovation can emerge from team collaboration. And individually, we can experiment with creative freedom by giving ourselves the time and space to express ourselves. As the saying goes, "The best way to predict the future is to create it."

30

Recapping WorkHuman

In Episode 084 Recapping WorkHuman, co-host Angie Cross quoted the wise warrior of the human spirit, Meg Wheatley: "Very great change starts from very small conversations, held among people who care."

This was our experience at the WorkHuman 2018 conference, where we interviewed Jennifer Fondrevey, a chief humanity officer who specializes in mergers and acquisitions and seeks to rediscover the humanity in business. Her personal experience working through mergers in her career inspired her to make a difference because, as she said, "Oftentimes, the human side is lost during mergers and acquisitions."

We also talked with Angela Romei, director at Panorama and leader of their paid leave program. Romei discussed why paid leave is a requirement for a more human workplace. She said, "If we want somebody to bring their whole self to work, we need to support their whole self, which means saying we understand that you need to take some time off, we don't want you to worry about your paycheck." Additionally, Romei explored with us many of the unintended consequences that can arise by not providing paid leave and shared several resources available on the Panorama website for leaders to plan and execute paid leave programs for their employees.

31

How to Cultivate Happiness at Work

As hosts of *Human Current*, we were frequently asked how we practiced what we talked about in our podcast. In response to these questions, here are some of our practices:

Find the balance. Finding balance can be thought of in so many different ways; for example, balancing work and personal life, family and professional relationships, fun and being serious, diet and fitness. Although our team usually felt in balance, like any healthy relationship, it took work. We often used nature to find balance and cultivate happiness. Going out to nature helped us re-energize and get grounded. Nature also provided us with opportunities to recognize patterns and boundaries since other ecosystems, like organizations and communities, often mimic nature.

Allow for emergence. While our podcast's success was a result of collaboration and hard work, we also trusted the process and allowed for emergence. We practiced this by reflecting on our experiences, challenges, opportunities, and the direction we were headed, along with recognizing the complexities present in those various situations. Allowing for emergence also meant that we stayed present and open, ready to adapt.

Understand and discuss values. We understood our own values, each other's values, and *Human Current's* values. We had discussions about how our values were similar, different, and competing. We expected and appreciated that we brought our whole selves to work, which included our values. We recognized that similar values can play out differently for all of us and that values shift and change over time; so, ongoing discussions were essential because this was not a one-time conversation. We learned that discussing our values reduced judgment and increased empathy.

Be honest and vulnerable. Over the years, we developed a trusting relationship, which allowed us to be honest and vulnerable with each other. We know that's easier said than done. We asked questions, of ourselves and each other, and held space for courageous conversations.

We didn't always agree with each other, but we trusted each other and had a deep level of respect for each other.

Give cheers. On those work days that went into evenings, we poured some wine, gave cheers to something we were grateful for, rolled up our sleeves, and dove into our collaborative projects or recorded a podcast episode. Giving cheers was also about celebrating. We celebrated our big and small achievements, milestones, ah-ha moments, and even painful learning. Giving cheers was one of the ways we "worked happy."

Foster an open system. We treated *Human Current* as an open, living ecosystem by exchanging information and using cognition to adapt, learn, and grow.

Be generous. While generosity is often associated with giving money, we practiced generosity with words, energy, attention, stories, and knowledge. The best generosity is the kind that doesn't come with ego or an IOU. Generosity can be contagious and spread throughout a complex system.

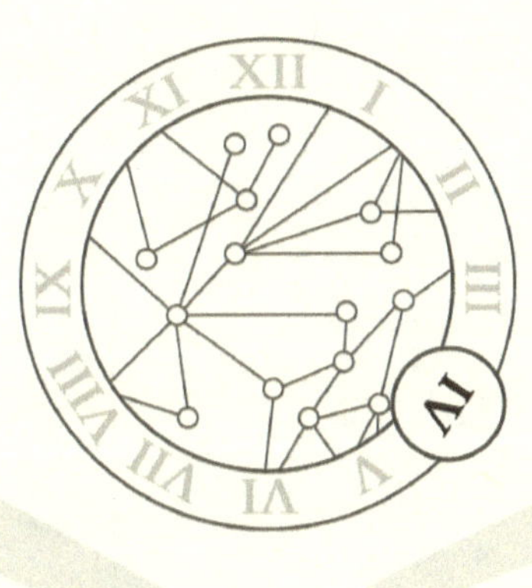

APRIL

1

Appreciating Nature

From the very first time we met Doug, he taught us that complexity is everywhere, and especially in nature. He would often tell us to go to nature for answers. One of the ways we would practice going to nature was by doing "walk and talk" instead of "sit and tell" meetings. When we were on our walks, he'd frequently insist on stopping to notice how the flowers or leaves on a tree looked up close. He'd point out fractal patterns in nature and discuss how they represented patterns in life and business. Through this process, he taught us to take our observations from nature and apply them to business.

Sometimes looking to nature was not about learning, but rather about being present and appreciative. For example, Doug shared with us that though his Dad had passed many years ago, he would still come around as a red cardinal to say, "Hi." We loved to witness times when a cardinal would all of a sudden perch itself on the windowsill of his office, and Doug would take a moment to recognize that his Dad just popped by for a visit. This made such an impression on us that we began pausing to say, "Hello!" to Doug's Dad when a cardinal would perch on our office windowsills.

2

This process of adapting as you go is what
we humans have completely lost sense about,
we just plunge ahead,
but every other living system
on the planet uses its intelligence,
uses its sensing to take in information
from the environment.
They have a fundamental freedom
in that they can decide what to notice,
and then they can decide how to respond
to what they've noticed.

Meg Wheatley
Episode 090 Reclaiming Leadership for the Human Spirit

3

Learning from Nature

In Episode 051 Learning from Nature, we talked with public speaker, author, and education transformer Laurie Marshall. Marshall summed up her work with, "I want to use my genius to fix what breaks my heart," and she posed the question, "How does nature solve a human challenge in the local ecosystem?" Marshall discussed her passion for empowering youth through creative collaboration and shared the importance of looking to nature as a tool for learning and a way for everyone to discover their inner genius. "Go outside and learn about your local ecosystem! And walk outside with a young person," said Marshall.

4

Collect Feedback

If a company wants to "manage" the health and well-being of their employees, leaders must collect and study feedback. Feedback is crucial to sustainability; however, simply collecting it is not enough. Leaders need to recognize that people have different rhythms and timescales, so sometimes observation is better than intervention. Not all feedback requires a response, but it is still important to constantly collect and analyze it, keeping in mind that the goal is long-term sustainability.

5

Assessing Culture in Complex Systems

by Sonja Blignaut for *Human Current*

One of the key shifts required in organizations is how we view culture and culture transformation. Executives have realized the importance of culture. "Culture eats strategy for breakfast" has become one of the most-used phrases around. Another often-used phrase is, "Soft stuff has become the hard stuff," meaning that executives realize culture can no longer be seen as unimportant and something that can be delegated to HR. Even with this new realization of importance, and all the various culture models and culture change tools around, sustainable cultural transformation still seems to be elusive in most organizations.

I believe that a big part of the problem lies in the way we think about culture. In most of the organizations I work with, culture is still seen as something that can be accurately diagnosed and designed. There is a belief that the leadership team, or a group of expert consultants, can design and implement an objectively ideal culture. The issue with this is that culture is an emergent property of a complex system. It emerges and evolves over time from a myriad of interactions between people, their environment, and the structural aspects of the organization (e.g. processes, policies). It often cannot be fully articulated, but it's known by everyone who is a part of the system.

Because the culture is ever emerging and evolving, any attempt to accurately "diagnose" it is flawed. Any culture audit result is only ever a partial point-in-time description, and the very act of diagnosing changes the system. The way we go about the diagnosis is also problematic. From a complex systems perspective, typical culture audit instruments have several shortcomings:

They are evaluative, not descriptive. A typical culture audit instrument asks people to evaluate the various aspects of the culture. For example, staff might be asked to evaluate their leaders in terms of whether or not they are appropriate role models to follow, or whether they live the organization's values. The problem with this is that an evaluative process is always open to gaming. Evaluations and opinions are useful, but they are highly subjective. It is much more useful to have a rich description of the experiences or observations that inform these opinions, because those are specific enough to act on.

Questions lack context. In most of the surveys I've completed, I've wanted to answer: "It depends . . . sometimes I feel my leader is a good role model, sometimes he isn't. Sometimes I feel he lives the values, sometimes he doesn't." This leads to a "middle of the road" answer that has very little utility.

They are based on a specific model or hypothesis. Most culture auditing tools are based on someone's model of what culture is and what the ideal looks like. Anything that falls outside of that framework will not be represented in the output.

Because culture emerges from our interactions and the stories we tell about them, one way to get below the surface and tap into this ongoing conversation is by continually gathering micro-narratives and observations from across the organization. The key is to ask people to *describe* the system, not *evaluate* it.

Once we have a richly described cultural landscape, we are able to evolve the system forward based on patterns across multiple observations. Some of these patterns may lead to large-scale interventions that span the organization (e.g., changing the approach to performance management). However the data also allows us to evolve forward based on small local actions aligned to a coherent direction. This is done by decision-makers across all levels of the organization by accessing the stories and observations from their area, and then asking a simple question: what do I need to change or do differently to create fewer observations like *these*, and more like *those*?

This is a question that someone on a board level understands as well as someone on the shop floor. It allows the entire company to evolve the culture into a coherent direction, but with local autonomy in terms of how they do it. It's also an un-gameable process.

The only way to change people's observations or experience is to actually change something in the environment or in the way they interact. Notice that we don't design an ideal culture for a potential future state that might never come about. We evolve forward from where we are today and learn and adapt as we go. This is a big mindset shift—for leaders and practitioners alike—that challenges our need for control and our need for the plans and linear processes we are so comfortable with.

Excerpted from a blog previously published at human-current.com/blog on September 26, 2016.

6

Reductionism got us to where we are and it's done a lot of good things, but it has also created a lot of problems, so we really need a new way of thinking.

Joss Colchester
Episode 089 Nonlinear Systems and Technology with Complexity Labs

7

Work-Life Balance

Our *Human Current* team could definitely vouch for the importance of work-life balance. Our office was family-friendly, we practiced flexible work space and work hours, and we encouraged each other to take care of ourselves and our families first. Ultimately this allowed us to perform our best at work. We were a small team of women—wives and mothers—so we greatly appreciated the tolerance for life interruptions. It was a careful balance of holding each other accountable, and giving grace when needed.

8

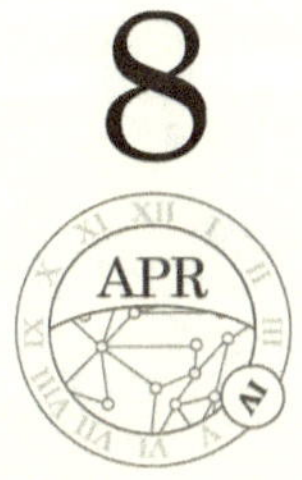

Planning for Emergencies

Emergency situations are a great example of complex systems at work. They are nonlinear, have an order-chaos dynamic, and are characterized by emergence. Emergencies create fascinating self-organizing systems. They are systems which are fully liberated to adapt to their environment. When we talked with Thomas Appleyard, manager of planning and programs with Ontario's Ministry of Health and Long-Term Care for Episode 018 Preparing for Complexity in Emergency Management, he said, "A self-organizing system is coordinated, but doesn't have a coordinator."

A crisis is an especially challenging self-organizing system to navigate because it emerges straight out of chaos, making adaptability a crucial element for reaching order. The real challenge is planning for adaptability. We all hope to have a plan and feel prepared for emergencies, but how do you plan for the unpredictable?

There are no absolutes in preparing for crisis, but there is hope in planning for a non-plan. Like a democracy, the strength of a self-organizing system is derived from its tolerance, adaptability, and resilience. Planning for a crisis requires fostering systemic resilience. By continually communicating ways to understand and deal with unpredictable systems, people can help each other become more comfortable with the idea that there is no plan—only tolerance, resilience, and preparation.

9

Connections Are More Important than Leadership

Establishing and empowering networks can make a huge difference during an emergency. This requires building relationships and figuring out the most effective ways to stay connected. In a presentation given at the HICSS-42 conference, "Complex Adaptive Systems: Emergence and Self-Organization," Stephen Kaisler and Gregory Madey discussed disaster management and emergency response and the importance of relationships between agents. "The relationships between the agents are generally more important than the agents themselves," stated Kaisler and Madey. Networks need strong communication channels and decentralized decision-making. These factors allow for optimal sharing of situational awareness and real-time data, as well as better coordination for backup support.

Emergency situations rely heavily on networks, but leadership can also play a role. Depending on the circumstances, leaders can be more or less influential due to the fast-paced and complex nature of each situation. Imagine a flock of geese flying in an organized V-shape or a school of fish swimming in unison. Do you think these systems are managed by one particular leader? Probably not—and these systems are generally flawless.

Human systems during an emergency are chaotic, and all of the forcing functions are so quick and emotionally charged that acting leaders, or in-the-moment leadership, can emerge out of this chaos. Therefore, those who project calm, are empathetic, and demonstrate resolve may have others looking to them for guidance and purpose.

10

Anticipate Emergence

Emergency situations require handling and even embracing ambiguity. Emergence is an inevitable factor—Thomas Appleyard pointed out that the word "emergency" has the same root as the word "emergence"—so attempting to place too much structure or too many demands on the system will not work. Stephen Kaisler and Gregory Madey described emergence as, "Rather than being planned or controlled, the agents in the system interact in apparently random ways."

Decentralization works best—specifically, the decentralization of knowledge, information, and communication. A flexible framework, which allows for distributed decision-making, is the best way to enable agents and networks. It gives them the authority to change their behavior in response to changes in the environment. Everyone within the system should be empowered to act, intervene, and even improvise if necessary.

11

Prepare with Storytelling and Reflection

In a crisis or emergency, use stories to examine complex problems and learn from past experiences. Then ask questions. What went right? What went wrong? What were the gaps? Were there any patterns that emerged? Were there any leverage points, or points where a simple intervention could have caused a lasting, direct effect? This process should help people explore, recognize, define, and communicate the elements which might arise during the next crisis situation.

Provide a productive way for people to communicate outside of chaos. Feedback loops will occur quickly during a chaotic event, so they should be addressed and reflected upon at a later time, in a calm environment. At that time, you can take Jon Kolko's advice from "Design Thinking Comes of Age" for *Harvard Business Review*, "Create models to examine complex problems. They add a fluid dimension to the exploration of complexity, allowing for nonlinear thought when tackling nonlinear problems."

12

Humanity is globally synchronizing
to the rhythms in the earth's magnetic field.

Rollin McCraty
Episode 073 Coherence and the Heart's Intuitive Guidance

13

Developing Innovative Ecological Systems

We talked with Chris Ling for Episode 052 Developing Innovative Ecological Systems. Ling is the director of the School of Environment and Sustainability at Royal Roads University. He teaches sustainable development, land use, environmental planning, and research methods, and has a particular interest in "a variety of interfaces between human and ecological systems." He also expressed an interest in climate change; however, not in the typical way of looking at adaptation and mitigation. Instead, he is interested in how "urban biodiversity interacts with climate change, impacts climate change, and can actually help communities with managing the effects of climate change."

Ling discussed how society tries to separate human and ecological systems to manage them, which he said is like "trying to pretend that urban areas are not places for ecology." Ling said, "This is being manifested in the way we managed our environments over the last few generations, where you push ecology off into the wilderness, into the remote places, into the places which you can safely ignore if you live in the urban area." This is part of the problem in "maintaining biodiversity, and maintaining healthy and resilient urban systems as much as ecological systems."

Ling also shared his experience with sustainable community planning throughout his career, including some fascinating research he conducted on the scalability of farmers' markets. "Our research indicates that once the farmers' market gets over a certain size, either in terms of the number of vendors, or customers, or the diversity of goods, then the localness of the markets can get diminished," revealed Ling. "The situation locally needs to be the thing that drives the solution."

14

Instead of being someone who is for fossil fuel or against fossil fuel, what if we change the story to: How do we create clean renewable energy because the next generation's survival depends on it? [Then] we can come up with new ways. It's an exciting time to come up with the third way, the new story, instead of just looking at people as our enemies.

Laurie Marshall
Episode 051 Learning from Nature

15

Rethinking the Story of Separation

How we think about progress and innovation needs to evolve beyond the story of separation. In an interview with Michael Wayne for *Interviews with the Leading Edge*, Charles Eisenstein said, "There are so many habits that we have of separation, and that means a really profound healing process needs to happen."

We can heal ourselves and our planet. We can look to nature as our mentor, and we can feel inspired by the world around us. Biomimicry can help us learn how to redesign our lives and our communities in a way that is supportive of *all* life. We have learned that sustainability should really be more than a separate field of expertise or an industry. Rather than a goal we work toward, it should be an integral part of our everyday lives. As we learn to use integral thinking, our mental models of dependence on the economy, first and foremost, will begin to shift. The story of separation confines us to short-term thinking. We can, and must, retell this story.

16

To divorce ourselves from nature
and use it as a commodity
has resulted in sickness and alienation,
the warming of the atmosphere,
the acidification of the oceans, and
the sixth greatest mass extinction event in history
that's currently underway. It's a great sickness
in our human population around the world.

Laurie Marshall
Episode 051 Learning from Nature

17

Reinventing Global Security

Why are all our biggest global problems caused by reductionist thinking? How do the environment and sustainability influence global security? In Episode 054 Reversing Complex Problems for Global Security, we explored these questions with global security thought leader James Greyson.

Greyson discussed research he conducted for the North Atlantic Treaty Organization (NATO) Science Programme, which is documented in *Seven Policy Switches for Global Security*. Each of the seven policy switches outlined in his paper offers an expanded vision of people's role on Earth and a whole-system change to implement it. The strategy he presented synthesized several major problems into one global issue of security, and he did this by pinpointing patterns within the global system. For example, he said, "Climate change is a perfect example of a global problem that we failed to solve by staying stuck in various kinds of default answers that either avoid the complexity of the problem, or they avoid managing the complexity that causes the problems."

Greyson believes we can use the size of global problems to our advantage by looking at patterns and intervening in the system at leverage points which hold the power to change global behavior. He presented a strategy to integrate complexity awareness and applied systems thinking in order to reach a new era of policy making, stating, "This global security ambition can be sought with a new era of policy making that encompasses the indivisibility, scale, and urgency of all planet crunch issues." Greyson's term, "planet crunch," is used to describe an apex to the plethora of global issues which threaten our global security, including: unequal access to energy, water, food, natural resources, funds, cooperation, as well as world-wide susceptibility to infectious diseases, climate instability, and ecological collapses.

He summed up, "I think global security could be an ideal way to describe an overall goal for complex, interlocked, global problems, because you've got so many problems which you can lose track of individually if you just start making a list. To actually see the problems as a whole set, it can help to have an overall goal."

18

There is no real boundary
between an organism and its environment.
It's just a constant energy flow,
and it's a constant exchange of information,
and the organism adapts.

Meg Wheatley
Episode 090 Reclaiming Leadership for the Human Spirit

19

Localism and the Economics of Happiness

In Episode 091 Localism and the Economics of Happiness, we had a conversation with author and filmmaker Helena Norberg-Hodge. We learned that Norberg-Hodge was a pioneer of the "new economy" movement and has been promoting an economics of personal, social, and ecological well-being for more than thirty years. She said, "Our dominant economic system is actually anti-life. It is imposing a monoculture both in biological terms and in human terms. A consumer monoculture is being pushed around the world." Norberg-Hodge discussed the harmful consequences of economic globalization and advocated for localism as a systemic path to return us to "the reality of complexity, diversity, and interdependence of all life."

Norberg-Hodge also shared the importance of spreading systemic and economic literacy, which is a mission she works toward at the organization she founded, Local Futures. She passionately believes "localization is about a deep reconnection to others and to nature."

20

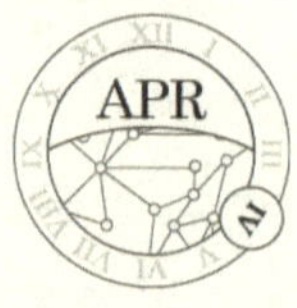

Competition in nature is always in the context of an overriding, interconnected mutualism where all organisms benefit from keeping a balance of eating and being eaten.

Laurie Marshall
Episode 051 Learning from Nature

21

Gaia Theory, Holistic Science, and Deep Ecology

In Episode 112 Gaia Theory, Holistic Science, and Deep Ecology, we had a conversation with Stephan Harding, an ecologist, author, and founding member of Schumacher College. Harding discussed his passion for the Earth and why he believes so deeply in helping others feel more connected to our planet. "What makes me passionate is the Earth herself . . . and so my work is to share that passion with people using a mixture of science, music, poetry, psychology, more or less any means I can to help people develop a deep sense of connectedness with the Earth."

Harding described how the Holistic Science course at Schumacher College offers unique learning experiences to help students connect with and learn from nature by integrating complexity theory, Gaia theory, and deep ecology. He claimed, "The way Schumacher College works is not like a regular college. It's based on the inspiration from India that when you come to learn you should learn with your head, your heart, and your hands. It's like a Western-style ashram-inspired learning community."

Harding also talked about how more expansive approaches to science, which integrate diverse methods of inquiry beyond just reason, can help us better understand the implications of complex problems, like climate change. He said, "Holistic science is really a journey of the soul toward wholeness, involving thinking, feeling, sensing, and intuition."

22

Becoming Environmentally Conscious

In the spirit of Earth Day, we got behind the mics for Episode 053 Becoming Environmentally Conscious in a Complex World to share clips from past guests Laurie Marshall, Chris Ling, and James Greyson. We asked them to respond to the following quote by environmental activist Annie Leonard: "The assumptions that 'pollution is the price of progress,' or that 'we must choose between jobs and the environment,' have long limited our creative thinking about innovative solutions that can be good for the environment, the workers, and a healthy economy."

Marshall, who is an author, public speaker, and education transformer, shared that we can use nature as a guideline to build innovative solutions which could change our world. She referenced examples of biomimicry, an approach to innovation that seeks sustainable solutions to human challenges by emulating nature's time-tested patterns and strategies.

Ling, director of the School of Environment and Sustainability at Royal Roads University, shared a perceptual dilemma about how we think about the environment as some "thing" that is away from us and outside of us. He explained that because many people live in cities, they feel separated from nature.

He also referenced an old model on sustainability which contains three circles arranged in a Venn-type diagram—the economy, society, and the environment—and according to this outdated model, where they overlap in the center is where we can achieve sustainability. He pointed out that this reductionist approach to sustainability is flawed because it implies that the economy can exist independently of society and the environment when, in fact, there would be no society without the environment, and there would be no economy without society.

Ling explained that a more realistic and accurate model shows the nested dependency relationships, placing the economy as the smallest circle within a slightly larger circle that indicates all of society, which itself is contained within the largest circle, the environment.

Greyson, global security thought leader, explained that our way of living is revealing a systemic error that drives acute problems. Similarly to Annie Leonard, Greyson strongly advocates for a circular economy, which he described as "a design correction to markets or capitalism as a whole to ensure that resources of all kinds end up again as new resources for people or for nature."

23

I often say localism, as a systemic path,
is an ism that can end all isms
because it is a return to the reality of
complexity, diversity, and interdependence of all life.

Helena Norberg-Hodge
Episode 091 Localism and the Economics of Happiness

We Are Complex Adaptive Systems

A powerful starting place and opportunity to understand complexity is within ourselves, since we are a complex adaptive system. Our body and mind are comprised of millions of cells that interact and connect with each other in non-linear and dynamic ways to produce behaviors and functions that emerge and cannot be understood by looking at one cell in isolation. Our bodies are comprised of biological interconnections, relationships, processes, and systems that scale to co-create life and consciousness.

There is something very deeply amiss with our Western worldview, so what we [Schumacher College] are trying to do with holistic science is try to understand what went wrong with the Western worldview and why it went wrong and then to create an expanded Western worldview. We don't want to reject all the great advances of the West, but we want to expand it so that it's no longer taboo to explore the psyche or soul through science.

Stephan Harding
Episode 112 Gaia Theory, Holistic Science, and Deep Ecology

26

Interconnections and Paradoxes

We cannot truly grasp complexity without a foundational understanding that interconnections are the fabric of all systems and all life. Even when we think about ourselves as independent agents with our own identity, thoughts, and behaviors, the significance of interconnection remains, for we are entwined with our surroundings and inescapably influenced by our environment. Separation is merely an illusion, created by our need to break things down in order to explain and understand them. As Alan Watts wrote in *What is Reality?*, "Black implies white, self implies other, life implies death." Thus, separation implies connection. Our society has not learned how to talk about the paradoxes that result from infinite interconnectedness. It's a conversation worth having.

27

Complexity Is a Young Field

The late Stephen Hawking said, "I think the next century will be the century of complexity." As a lifelong scientist, seeker, and questioner, Hawking obtained a profound appreciation for the ways in which science can influence how we see the world. His unfailing curiosity propelled his research endeavors and vision for a unified scientific theory about the mysteries of our universe. He worked tirelessly on his research, writing more than 140 papers in search of a unifying scientific theme that would bring people together and inspire conversations around science and why we exist.

In *A Brief History of Time* he wrote, "If we do discover a complete theory, it should, in time, be understandable in broad principle by everyone, not just a few scientists. Then we shall all, philosophers, scientists, and just ordinary people, be able to take part in the discussion of the question of why it is that we and the universe exist."

Hawking was a scientific pioneer who helped propel new discoveries and theories, and toward the end of his career he saw complexity as an opportunity for the future of science and the human race. Luckily, during his many years of research, there were scientists all around the world, including many physicists and mathematicians, who dedicated their time to research complex adaptive systems. For example, Yaneer Bar-Yam, a physicist and president of the New England Complex Systems Institute, has been contributing to research, education, and real-world applications of complex systems science since the late 1980s. Around the same time, the Santa Fe Institute became the first research institute dedicated to the study of complex adaptive systems. In many ways, we are just beginning to study and learn through a complexity lens. What a great beginning it has been.

28

I really want to see complexity science embrace multi-disciplinary participants, again because of its universality. It certainly applies everywhere.

Tanuja Prasad
Episode 118 Practical Applications for Interconnected, Nonlinear Systems

29

Complexity Is an Ancient Field

Although the scientific field of complexity matured late in the twentieth century and is young compared to many other scientific disciplines, its fundamental principles can be traced back to ancient philosophers, like Aristotle. There are also cultural ways of thought and many ancient wisdom traditions that share common ideas and language with complexity as a science and worldview. With a rich background as a young scientific field rooted in ancient principles and wisdom, there are many lessons we can learn from complexity.

It becomes extremely difficult to take a position
when you are seeing things as separate linear issues.
But if you step back you see that we have
these two systemic directions,
one of them ever more energy-intensive,
techno-intensive, capital-intensive,
taking us farther and farther away
from the reality of our dependence on
the living Earth, on mother Gaia.
And on the other hand, we have another path,
which is actually a multitude of paths
and so there is no one single formula,
there is no one imposition.

Helena Norberg-Hodge
Episode 091 Localism and the Economics of Happiness

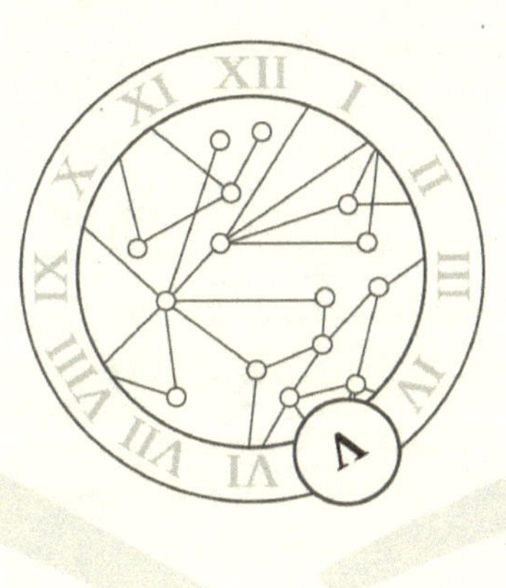

MAY

1

Lessons of Generosity

Human Current is a testament to what can be achieved with a little generosity. Doug gave us the mission to change the world by sharing complexity thinking with everyone around the world. With his generous funding of *Human Current*, we were able to make his vision possible. Doug even told us once, "Your budget is to change the world." We experienced Doug as the most generous man we have ever met, and not simply because he funded *Human Current*. More important than dollars, we benefited from Doug's time, compassion, and wisdom.

In the office, Doug was always up for meeting with us or grabbing lunch so we could catch up, ask him questions, and share ideas. Sometimes while we were out for lunch he would notice fundraisers, like paintings by school children hanging for sale on a restaurant wall, and he was always willing to give. Another example of his generosity is providing tablets and computers for schools in need.

Doug kept an impressive book collection at the office, which covered every topic under the moon. When he found a new book he loved, Doug would insist that several copies of the book be ordered so we could all have copies, and he would encourage us and others to take books from his bookshelves. Doug's love for books was a primary motivation for writing this book.

Doug's generosity and love extended to animals and the planet. For example, when the land behind the office was being developed and the deer were being pushed out of their habitat, Doug got a deer feeder to ensure they had food. Doug would also insist that the building lawn service was dedicated to *caring* for the grass and trees, not just *mowing* the lawn or trimming branches. He had a great appreciation for trees that existed decades, or even centuries, before we occupied the land.

These are just some of the ways that Doug taught us about generosity. We have learned that generosity is about giving time and wisdom. It's about being present and aware of opportunities to give, not just giving when asked. Generosity is sharing knowledge, giving encouragement and appreciation, loving, and giving without expectation or resentment. Generosity can be extended to people of all ages and walks of life, animals, and the earth. These experiences and lessons have changed us and we will carry these lessons of generosity with us for the rest of our lives.

2

I really believe humans are the greatest resource
for changing the world. It is up to us
to be there for each other, our communities,
and even people that we don't know very well.

D'Artagnan Caliman
Episode 034 Network Challenges Within Child Welfare

3

Design-Led System Transformation

We talked with whole systems design practitioner and author Daniel Wahl for Episode 122 Design-Led System Transformation. Wahl shared stories from his journey learning about complexity science, holistic science, and whole systems health. He described his Schumacher College experience, where he obtained a master of science degree in holistic science and was inspired to study natural design. He later received a doctorate in natural design at the University of Dundee. Wahl said, "For me design is human intention expressed through interactions and relationships. It touches almost everything." He also expressed his appreciation for biomimicry and the importance of learning from nature. "The high level of biomimicry and ecomimicry is understanding that we are biological beings, first and foremost, and that if life creates the conditions conducive to life as a general pattern, then we had better do our very best to learn from the community of life how we can meet our needs and become stewards and caretakers of the ecosystems we inhabit."

4

Dynamic Complexity and Systemic Structures

Complexity science helps us understand the dynamic, complex nature of systems, including how and why they emerge and co-evolve in certain ways. Dynamic complexity occurs when "cause and effect" are not close in time and space, and human intervention often results in unintended consequences. In these dynamic environments, seeing the underlying structure, context, relationships, patterns, and history are all key to understanding why certain individual and system behaviors occur. As Donella Meadows advised in *Thinking in Systems*, "Locate responsibility in the system. That's a guideline both for analysis and design. In analysis, it means looking for the ways the system creates its own behavior."

Systemic structures are powerful forces which shape key interconnections and ultimately influence patterns of behavior over time. Some structures within a network might limit access to information or resources for someone while reinforcing access for someone else. In network science, a subfield of complexity science that studies the structure of complex systems, this positive feedback cycle is called preferential attachment. Oftentimes, seeing these structures in our society is not easy because they are very subtle, nuanced, and deeply interconnected. The influencing structure is often not as obvious as a less connected node in a network, for example. In human systems, there are many hidden structures, such as policies, perceptions, or social norms, that influence how people make decisions and act. Looking for and studying patterns can help us identify many of the influencing structures within our human systems and society at large.

5

Stories Are Network Glue

Relatable human stories have the ability to bind a network together, especially when there is no formal structure within the network. It is important to note that the most powerful stories—meaning the ones that glue a network together—always focus on people, not ideas. They are relatable and human. These stories ultimately allow networks to quickly create order out of chaos because they promote a superior level of trust between agents in the network. They shape how we perceive our history and influence what we believe for our future. When people share common stories, they share a mindset. They create a sense of "us" which is the key to effective self-organization.

As Jonah Sachs wrote in *Winning the Story Wars*, "People who tell and believe the same stories hold the same values. They share a worldview. In fact, a worldview is just a collection of stories about how things have come to be the way they are and what we should do about them. A good story is therefore the fundamental ingredient in allowing humans to create a sense of 'us.' Shared stories, shared values, shared worldview—us."

6

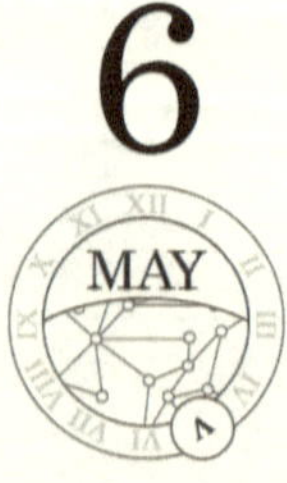

If we accept complex adaptive systems,
we are faced with a different set of questions:
How do we measure the impact of an action
in a world where everything influences everything else
because of interconnectedness?
How do we make decisions and choices in such a world?
How do we design solutions and actions
when phenomena are nonlinear and have tipping points?

Tanuja Prasad
Episode 118 Practical Applications for Interconnected, Nonlinear Systems

7

Exploring Complex Human Networks

In Episode 032 Exploring Complex Human Networks, we talked about emergence and self-organization within human networks, using Foster Care Alumni of America (FCAA) as an example. FCAA is a nonprofit organization dedicated to connecting and empowering adults who were once in foster care. It's an amazing organization because it not only provides a sense of belonging and support for those who have gone without the privilege of family—our most core human network—it also inspires action and reform within systems of child welfare. FCAA's tagline is "connecting today, transforming tomorrow," which speaks directly to their mission to support a community of people with similar life stories and inspire hope and drive change within the entire foster care system.

The power and strength of human networks becomes very apparent when you take a closer look at organizations like FCAA. Despite FCAA's limited funding and tiny organizational structure compared to other organizations of its scale, FCAA's network has continued to grow larger and stronger each year. FCAA is an example of an emergent social network, powered by shared narratives and shared values. It's the collection of members' stories, along with their shared values, which propels network growth in organizations like FCAA.

8

I see the complexity worldview as really underpinning and giving us some approaches to look at issues of social justice, inequality, climate change, and economic and political systems, as well as how to manage organizations.

Jean Boulton
Episode 075 Embracing Complexity

9

Human Stories for Network Change

In Episode 033 Human Stories for Network Change, we interviewed Jasper Faolan, a writer, psychiatric nurse practitioner, and founder of Journal to Save Your Life (J2SYL), a free online mental health program. She advocates self-expression, like storytelling, as a way for people to heal from trauma and mental health problems.

Faolan explained how J2SYL promotes social change by connecting the dots within complex adaptive systems: "A lot of times people come to me and say, oh you're doing this really neat thing, what would you like me to do for you? And I like to reverse it, to be able to ask a person to show up in exactly the way that they would like to and in the way that their life has intended. That is where the gold is and that's where you get to connect people around this higher mission you're all working toward. I think the biggest thing is to be a listener and a connector." She says that storytelling works because, "There's this tipping point between that initial assessment and talking with a person and listening to their story, into moving them into a place where they see their empowerment, their incredible beauty, and the way that they're able to uniquely contribute to the world."

10

We keep trying to change the overall culture,
which is always an emergent phenomenon,
by working backwards, by changing the parts.
But it doesn't work that way.
Life doesn't change that way.

Meg Wheatley
Episode 090 Reclaiming Leadership for the Human Spirit

11

Network Challenges Within Child Welfare

For Episode 034 Network Challenges Within Child Welfare, we spoke with D'Artagnan Caliman. He posed the question, "Are we, as communities, thinking about what we can do as a human network to support families that are in need?" Caliman, chief of staff for the Maryland Department of Human Resources, discussed the complexity of serving two million people with many different interconnected network service groups and stakeholders. He said, "Our work is really focused on safety, permanency, and the well-being of children and families . . . ensuring that a child and family are holistically successful."

Caliman described the importance of cause prevention in the child welfare system, rather than simply treating the symptoms of family dysfunction as they arise. He said, "We have to reach out to our constituent groups to find what is working and build upon those successes, and also to hear from them on what is not working." Another quote that stood out from his interview was, "If we are unsuccessful at helping children and families to become stable and viable within their communities, then we are failing our communities and we are failing our children."

12

It is the human desire for a simple, single, static answer to things that creates a contrast between what we want to see and what exists. Once we get over that desire for simplicity or simplification, complexity science will become the fundamental assumption of all science going forward.

Tanuja Prasad
Episode 118 Practical Applications for Interconnected, Nonlinear Systems

13

Modeling and Forecasting

We were the official podcasters at the International Conference on Complex Systems 2018, which is where we talked to Irene Bosch and Elena Naumova for Episode 105 Modeling and Forecasting Infectious Diseases. Bosch is a research scientist at Massachusetts Institute of Technology who has worked with tropical diseases and viruses like Zika virus, dengue fever, and chikungunya.

Bosch discussed the lack of research and understanding of how diseases are diagnosed and spread. She stated, "We are facing a twenty-first century problem with tools that are from 100 years ago. We still die from a mosquito bite." Additionally, she described the potential benefits of interdisciplinary collaborations in forecasting epidemics and finding solutions for diagnoses. Bosch said, "We hope with solutions like modeling, forecasting, and taking into account environmental factors, movement, and land use, we will be able to predict hot spots or even predict when in the year these epidemics will occur. Then we will be able to intervene in a very early fashion."

Bosch and Naumova are colleagues. Naumova is a mathematician and director of the Tufts University Initiative for Forecasting and Modeling of Infectious Diseases. In the same interview, Naumova talked about how she applies mathematical models to better understand different kinds of biological phenomenon. She said, "My passion for a long time has been to learn how to apply mathematical models, computational models, and statistical models into something practical."

14

Systemic thinking is how we save the world, because the problems we are facing today are unprecedented.

Angela Montgomery
Episode 036 Harnessing System Potential

15

Understanding Unintended Consequences

In Episode 055 Unintended Consequences, Yaneer Bar-Yam shared with us how quickly unintended consequences can ripple throughout our global systems. He discussed the causes of increased global food prices from food crisis research he conducted at The New England Complex Systems Institute (NECSI).

Bar-Yam said that today almost 50 percent of the United States corn crop is used to produce ethanol, the substance that makes up 10 percent of gasoline. United States energy and environmental policy led to this shift in the use of biofuels for ethanol; however, unintended consequences soon emerged. Bar-Yam explained that corn is such an integral part of how we produce food because it provides feed for animals that provide meat and dairy products, and it is used in the production of sugar, among many other uses. So, reallocating such a large portion of a key crop for the production of ethanol was very disruptive to food prices on a global scale.

There was another factor that he found contributed to increasing global food prices—a deregulation of commodity markets. When the mortgage and stock markets crashed in 2008, money from those markets went into the commodity markets, including agricultural commodities and foods, which also contributed to rising food prices around the world.

The fact that these two seemingly straightforward changes created such disruption was an important discovery; it proved there is vulnerability in our global systems. The actions of one country can quickly spread to influence another, and in impactful and unanticipated ways. Bar-Yam's findings—that policy change and market deregulation contributed to a rise in global food prices—are significant because they identify a starting point for what could become global scale disruption, since food systems are so intertwined with human systems.

Bar-Yam made the profound connection from food prices to events like the Arab Spring and the breakdown of order in several countries, including Syria. He explained that civil unrest, revolutions, and the flight of refugees are all events which resulted from many civilians being unable to feed their families and feeling frustrated with unsympathetic governments. Bar-Yam stated, "We are watching this global cascade—that goes from economic factors, to food, to riots and revolutions, to refugees—cascading around the world."

Bar-Yam emphasized, "We need to understand global consequences in order to be able to act and react effectively to the challenges we are facing today." This describes the purpose of the research conducted at NECSI and justifies complexity science as a way to help us effectively pinpoint, navigate, and solve challenges in today's world.

16

If people can recognize the value
of allowing themselves to turn off the conscious brain
and let their subconscious do what it's designed to do,
then we would all be better off, or at least
we will understand our decisions
in a more real-world way.
We think we are rational beings,
and if we go on assuming that
and not really understanding
how we make decisions in our everyday lives,
then we actually act even more irrationally.

Spencer Gerrol
Episode 042 The Neuroscience Behind Your Vote

17

A Complex Systems Approach to Countering Mind Control

In Episode 094 A Complex Systems Approach to Countering Mind Control, we spoke with author, mental health counselor, and founding director of the Freedom of Mind Resource Center, Steven Hassan. We first met Hassan at the International Conference on Complex Systems 2018 where he spoke about cults and mind control as "a systematic social influence process."

Hassan spoke about his work and described how a complex systems model can be used to mobilize social networks, including policy makers, educators, health professionals, and law enforcement, to empower those impacted by undue influence. He said, "We're on one planet and the future of our species and every other species on the planet counts on us making ethical decisions based on our longevity. I worry about groups that preach Armageddon and somehow some magical mystery force is going to wipe the planet clean and only the true believers will survive. That's a very dangerous ideology. I think everyone needs to come together and realize we're humans, we have kids and grandkids, and we've got to take care of them."

18

We are exploring this
wonderfully new participatory worldview,
that fundamental shift from . . .
prediction and control of a system that we observe
as so-called "objective observers"
to a science that really understands
that we are always subjective participants
in the complex dynamic systems that we participate in,
and that the best science can do
is to shift from this prediction-and-control obsession
to aiming to guide appropriate participation.

Daniel Wahl
Episode 122 Design-Led System Transformation

19

What Japanese Art and Complex Adaptive Systems Have in Common

Kintsugi is the Japanese art of mending the cracks of broken pottery with gilded lacquer. We first heard about kintsugi from Doug years ago. We even have an image of kintsugi on our office fridge! So we were thrilled when, in Episode 003 Using Systems Thinking to Resolve Conflict, Jason Dykstra discussed what kintsugi and complexity have in common.

Kintsugi treats breakage and repair as part of the history of an object, rather than a flaw. Complex adaptive systems are like kintsugi in that they are history-dependent, which means that changes and even the "flaws" become part of the system. This is very similar to the philosophy of wabi-sabi, which embraces change, impermanence, and imperfection as part of the beauty of universal perfection.

I think it's important to actually create a distinction between self-management and self-organization. We find it's a really important and useful distinction. So, self-management is the ability of a team to help govern itself. Self-organizing is the ability for a team to pull together, form itself, disband itself, to have the creation ability innate to the team and not within a structure that lives above it or in a command-control system.

Bud Caddell
Episode 069 Designing Meaningful Work in Complex Times

Exploring Complex Human Networks

In Episode 032 Exploring Complex Human Networks, we learned about action research and introduced the topic of complex human networks with guest Mary Anne Herrick, the communication program officer at World Vision and president of Foster Care Alumni of America (FCAA). Herrick described her work at World Vision, "Our overarching vision is to bring clean water to every person and every area where we work by the year 2030, so within our lifetime."

Herrick described how providing clean water can impact a complex system, such as a community. She said, "When you bring clean water to a community, it improves health, it helps children be able to go to school, it frees up women to have time to do productive activities, it ultimately impacts all these different parts of their lives. So I see water as the foundation to impacting the complex system of humans. One of the most powerful ways that World Vision is impacting communities that need clean water is also by empowering the community members to advocate on their own behalf with their government to basically hold them accountable and demand the basic services that they should have as just a basic human right, like clean water."

This focus on advocacy creates a connection between Herrick's work with World Vision and with FCAA. She said, "Human networks and system impact are also in the work I do with Foster Care Alumni of America because there we're working with our peers and young people and adults from foster care, and empowering them to advocate with the government on their own behalf for their basic needs as well." Herrick discussed FCAA's mission to "connect and transform" the foster care system to impact the outcomes for those who are coming up in the system. She added, "As people who have been through the foster care system . . . we also advocate to improve the foster care system for the young people who are aging out and who are going through this system now."

Co-host Angie Cross shared that she was a co-founder of FCAA and her thesis project for her master of arts in leadership focused on FCAA's membership engagement. She worked closely with Herrick, using action research as one of her research methods. Marvin Weisbord described action research in *Productive Workplaces: Dignity, Meaning, and Community in the 21st Century*: "Action research became the opposite of scientific research. Instead of standing outside the experiment, watching what happens and writing up your findings, you become a learner in a situation you helped devise. You develop a stake. You assume responsibility."

Herrick was eager to participate in this action research project because it was consistent with her views. "My perspective on human networks is that we as humans are intimately connected to each other in profound ways that help us survive and help us get our needs met."

22

We are not becoming a homogeneous world society, we are actually becoming an integrated but differentiated system where parts are doing different things but are nevertheless strongly connected to each other.

Yaneer Bar-Yam
Episode 050 Complexity Science Is Everyone's Science

23

Redirecting Our Negativity Bias

What is negativity bias? Why are humans so prone to it? Negativity bias was first documented in 2001 by psychologists Roy F. Baumeister, Ellen Bratslavsky, Kathleen Vohs, and Catrin Finkenauer in the *Review of General Psychology*. Their article, "Bad is Stronger than Good," described the human tendency to be much more likely to be influenced by and to recall negative experiences, instead of neutral or positive experiences. It is a behavior that our ancestors developed in order to quickly identify and react to threats. It was a very useful fight-or-flight reflex.

Even though negativity bias was useful for our ancestors, who needed caution and skepticism for daily survival, can we say its overwhelming presence is serving us well in today's world? We would argue that, as a default mindset, it rarely serves us now. After all, human civilization has changed a lot and it continues to change very rapidly. Why do we continue to fall victim to a bias that served us so many years ago? Why are we stuck in this old way of thinking?

In today's world, there are many inputs which significantly influence our thoughts—like mainstream media and social media—and which thrive on our instinctive habit to focus on negativity. It's a matter of business for media, because we process negative data faster and more thoroughly than positive data, and it affects us longer. Negative news simply gets more attention than positive news.

In addition to this environmental influence, many of us "grow up in a world that doesn't value the training of attention or the capacity to cultivate specific emotions," as Tony Schwartz wrote for *The New York Times* in "Overcoming Your Negativity Bias." We all have the ability to decide what will earn our attention, and we can deliberately control and cultivate our minds to decipher our world with less bias. This deliberate action requires practice. It is very easy to fall back into old habits.

As agents within a collective human system, we adapt and interact in response to our environment, and our environment will continue to reinforce our negativity bias—if we let it. Consequently, it is easy to feel stuck in a feedback loop of negativity. We have a choice to make. We can either give in to bias and stay in the loop, or learn how to purposefully direct our attention to where it will serve us best. The liberating news for all of us is that we have a choice! If you decide that negativity isn't serving you, then you can redirect your focus. We can seek out and destroy our unconscious biases.

We can regain control of our minds and our everyday interactions in order to create new patterns of behavior. What's even more exciting is that as we change and evolve, so will our environment, because we are so interconnected within networks and have such a profound effect on each other. There is so much power in our collective consciousness.

We have the power to cultivate social optimism. As our individual thoughts, choices, and behaviors shift away from the default of negativity, change ripples through the system.

24

Life itself is an emergent property, so is consciousness. Why should we separate them with anything else?

Tanuja Prasad
Episode 118 Practical Applications for Interconnected, Nonlinear Systems

25

It's National Wine Day!

We love May 25 because in the United States, it's National Wine Day!

If you are a *Human Current* listener, you have probably heard us uncork bottles of vino and give cheers during many episodes. When we opened a bottle of wine and poured our first glass, we'd take a moment to acknowledge an accomplishment, person, or what we were grateful for that day, and then we'd give cheers to what we shared. Often we gave cheers to Doug, or to a guest. We'd take our first sip with purpose and, as our podcast guest Rick Hanson might say, "We took in the good."

We have felt incredibly blessed to have worked in such a special environment while co-hosting the *Human Current* podcast—often with a glass of wine in hand. Perhaps you've joined us by sipping on a glass of wine while you listened to an episode?

We believe sipping wine can be combined with many other activities to relish happy moments. Here are some of our favorite activities:

- Listen to the *Human Current* podcast or another podcast.
- Sit by the pool or on the beach.
- Draw, sketch, or doodle.
- Brainstorm ideas with colleagues.
- Take a bubble bath.
- Read a book. (Check out our Bookshelf at the back of this book!)
- Listen to music. Dance.
- Cook a fancy meal. Enjoy your fancy meal.
- Play cards or a board game with others.
- Watch a classic movie.
- Write a letter to someone special.
- With a friend, give cheers to an accomplishment or person.

You can find a list of the wines we have uncorked in our Wine Rack at the back of this book.

26

My favorite concepts in complexity would probably be ecosystems, patterns, and the concept of the part, the whole, and the greater whole. I think that they are so interrelated and at the same time give the most creative interpretation of how we live within and between complex adaptive systems.

Denise Easton
Episode 117 Influencing Pattern-Based Change

27

Navigating Systemic Risk and Critical Transitions

In Episode 119 Navigating Systemic Risk and Critical Transitions, we spoke with biophysical scientist and author of *Crashes, Crises, and Calamities*, Len Fisher. We found Fisher to be an often humorous storyteller. He claimed, "I'm not a complexity scientist, I've taken on a role as a proselytizer for complexity thinking."

Fisher gave many in-depth examples for how sudden change or critical transitions can happen within complex adaptive systems. "Systemic risk is the risk that a critical transition will occur in a complex adaptive system as a whole." In our discussion, he unpacked how interconnectedness, unintended consequences, runaway effects, and emergence all influence systemic collapse. He also shared what is needed to effectively govern complex systems. "If you are going to have effective governance of complex adaptive systems, then you need three things: you need to be watching for the warning signs, you need flexibility of your decision-making and ability to take action, and you need resilience or adaptation."

28

The discussion that needs to happen is not a fight, but an understanding of all the different perspectives within the spectrum.

Deborah Westphal
Episode 121 Preparing Leaders for Complex Change

29

Using Systems Science to Improve Community Health

For Episode 035 Using Systems Science to Improve Community Health, we talked with Daniel Taber, a scientist who specializes in food policy and systems research. Taber used real-world examples to explain the complex relationship between system science, science communication, policy change, and public health. He explained, "A key part of policy research is understanding how people and organizations adapt. And, of course, that's a key part of complex systems research—understanding how people adapt to a 'shock' in their environment. I think the key is understanding the systems that we live in, understanding those systems' goals, and strategically thinking about how we can utilize them better."

During our conversation, Taber shared that complexity thinking will give better results than traditional research. "Complex systems research is more focused on what we observe in the real world and trying to understand the dynamics that take place, rather than creating this artificial world. A lot of science—particularly health science, where I work—is built to reflect an artificial world that's very simple and unrealistic."

30

We can look at how effective reductionism is
to understand things . . . and at the same time
we can recognize that the reductionist approach to science
is really unable to understand any system in its entirety.

Jeremy Lent
Episode 074 The Systems Worldview of Human History

Practical Applications for Interconnected, Nonlinear Systems

For Episode 118 Practical Applications for Interconnected, Nonlinear Systems, we spoke with systems thinker and entrepreneur Tanuja Prasad. Prasad shared details about her relationship with complexity, including how it has shifted her perspective about life, work, and science. For Prasad, "Complexity science is a science of maturity." Prasad described the complex, nonlinear nature of systems and explained many practical concepts and applications for people working and living within systems. Prasad said, "Models are not reality, they are merely a window into reality and usually that window is defined by needs. So the real skill we need to develop is how to handle models, to switch them in and out as needed and to understand when one is not useful anymore. Basically, we need to be able to flow with models."

Prasad also expressed her passion for complexity science applications within the social impact sector and shared a quote by Peter Drucker, "The greatest danger in times of turbulence is not the turbulence itself; it is to act with yesterday's logic." Prasad said, "Complexity science is providing us with a logic with which to go forward, we just need to embrace it."

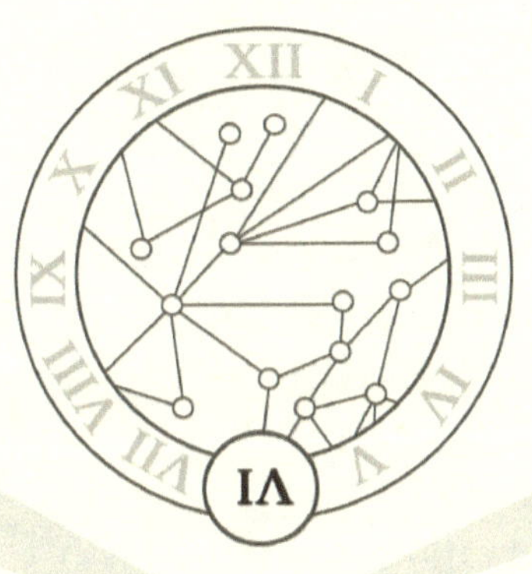

JUNE

1

Personal Wellness for Performance

Douglas outside his Austin, Texas office.

Doug taught us that personal well-being is a critical element in individual growth and effectiveness. He viewed three lifestyle factors as primary contributors to personal well-being: staying out of stress, dealing with negatives, and wellness.

Doug recognized that stress is counterproductive and should be avoided because going into stress not only lowers performance, it can also have a significant impact on physiological health and wellness. He would show us a chart, "The Performance Challenge Curve," to help us understand the point at which a healthy challenge can turn into unhealthy stress.

Doug recognized that negative thoughts affect our physical and cognitive systems detrimentally, pushing us out of total heart and mind coherence. He encouraged us to de-stress and look for the contrast the universe was presenting when negative thoughts would come up. "The Attitudes and Emotions Chart" was his tool for showing us when certain emotions might lead us to either connected or disconnected states. He wholeheartedly believed that there was good in the negative because knowing what you *don't* want ultimately helps you know what you *do* want. You just have to learn when to redirect your attitudes and emotions.

Finally, Doug emphasized that the mind-body connection is real and that wellness is necessary for both mental and social performance. He encouraged us to pay attention to what our physical bodies are telling us, to rest and recover when needed, and to enjoy fresh air, organic food, good water, sunshine, and exercise.

2

Most people think of consciousness as something that you have—like if my brain is awake I have consciousness—and I think of it more as something you tap into.

Julia Mossbridge
Episode 078 Loving AI

3

On Being Certain

In Episode 063 A Philosophical Look at the Mind, we shared our conversation with neurologist and author Robert Burton. We reached out to Burton after reading an article he wrote for *Aeon* digital magazine, "Our World Outsmarts Us." The article showcased his intellectual curiosity and humility as he addressed the idea that our human brains are perhaps more limited than we have assumed.

Burton argued that our social problems have become immensely complex, and that many of our current challenges result "in part from a basic lack of ability to sense how the present world works." His assessment on our mental limitations originally stemmed from his book, *On Being Certain: Believing You Are Right Even When You're Not*, in which he explored the paradoxical relationship between our thoughts and what we know.

Burton researched the concept of certainty with the use of medical data, experiments, and personal stories. His lifelong research led him to question many present-day assumptions about the human mind, and his discoveries on the feeling of certainty were especially fascinating. He described certainty as a mental sensation that happens to us, and that we cannot make the sensation of certainty happen.

Consider a time when you have had an ah-ha moment. Do you believe you made it happen? Or did that ah-ha come to you all of a sudden? This ah-ha feeling, which often occurs when something makes sense to us for the first time, is, in fact, our brain's reward system at work. Our brain rewards us with dopamine when we recognize and complete patterns. Burton explained that this mental sensation gives us the feeling of certainty, and it occurs even if we complete a pattern with inaccuracy. Our brain's pattern-recognition reward system activates even if we are wrong, simply because we have completed the task of creating meaning. Burton's research on our pattern-recognition system has identified an important limitation in our cognitive abilities.

Our recognition and appreciation for this very human limitation is sorely needed in society. As humans, we somehow have to resist our biological urge to feel certain, and instead embrace the very uncomfortable notion of uncertainty in order to navigate the complexity of the world around us.

So where does this leave us? Burton said, "The best defense against combative ideologies isn't more facts, but an admission of the limits to our knowledge." He argued that we need more tolerance, compassion, and humility to address our societal issues. Burton suggested that by admitting our limitations in the face of immense complexity, we could avoid counterproductive—and often unintended—consequences that arise out of arrogance and a certainty mindset. Tolerance and humility are needed more than ever to begin the tough conversations our society needs to propel forward.

4

The scientific worldview is essentially
a reductive materialist philosophy.
It says that if you want to understand something
you start taking it apart in little pieces,
and then you see how the pieces fit together,
and then you understand something. But as we know,
especially with your interest in complexity theory,
that when you start looking at complex systems,
you can take it apart into little pieces, but those little pieces
will not tell you how the whole thing is working,
and consciousness is a perfect example of that.

Dean Radin
Episode 081 The Science, History, and Implications of Real Magic

5

Public Health as a Complex System

Ever had trouble coming up with a single "right" answer? You're not alone. "Though we may be across the world, we breathe the same air," said public health policy scholar Michele Battle-Fisher. As we talked with her for Episode 013 Public Health as a Complex System, she revealed how systems thinking can bring new light to how disease, wellness, and the effects of policy spread through populations. She said, "I strongly believe that the innovation we need will come from bringing complexity and systems thinking into the discussion."

Our conversation covered everything from food deserts to "policy puffins," her shorthand reference to birds of a feather flocking together. We learned about the very real impact of chaos in health care, why time scales matter in measuring system impact, and why we need systems thinking education for tomorrow's policy makers. "Health is messy, health is complicated, and the elements that pertain to health are interdependent," said Battle-Fisher. She discussed "thinking outside the textbook," eureka moments, and radical learning experiences. Battle-Fisher also shared a Wayne Gretzky quote, "I skate to where the puck is going to be, not to where it has been," and noted, "To be a lone voice is difficult for systems thinkers."

6

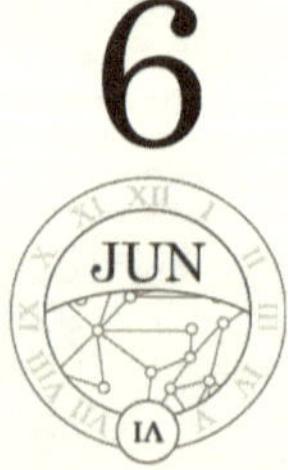

Wellness is Contagious

Where does wellness live? Is it in our minds, our health systems, or is it a complex system of shared responsibilities? How do you know when you've created a culture—and whose responsibility is it to create change? In Episode 014 Wellness is Contagious, we explored these questions and how the spread of information affects our health. Co-host Stacy Hale said, "We as individuals do have agency when we begin to see the systems we are a part of and broaden our mental models." This reminded co-host Angie Cross of a quote by John Sterman, "There are no side effects. Just effects." She went on to observe, "Like an epidemic, our wellness could also be contagious."

7

If we help people understand the whole system,
they will start to make, on their own,
some magnificent wellness changes.

Renee Moorefield
Episode 086 The Conscious Evolution of Leadership

8

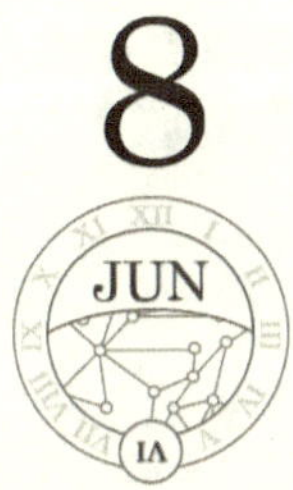

Mindfulness is Complexity's Best Friend

We talked with Pilar Angel, hypnotherapist and wellness coach, for Episode 026 Mindfulness is Complexity's Best Friend. Angel said, "The most complex system that exists is your own mind," and that if we "understand that you are an observer of your thoughts, your own complexity will become less mysterious." Angel shared tips for practicing mindfulness and explained how these practices can reduce stress and improve overall performance. She also quoted from Ellen Langer's article, "Mindfulness in the Age of Complexity" for *Harvard Business Review*, "Life consists only of moments, nothing more than that. So if you make the moment matter, it all matters. You can be mindful, you can be mindless. You can win, you can lose. The worst case is to be mindless and lose. So when you're doing anything, be mindful, notice new things, make it meaningful to you, and you'll prosper."

A Curious Metaphor of the Brain

In Episode 057 A Curious Conversation on the Brain and Body, we referenced Yaneer Bar-Yam's observations on the paradox of separation and integration in his book, *Making Things Work*. "Why is the brain subdivided at all? The reason is that when aspects of the world around us are partially independent, it is much better to store and act on them using partially independent parts of the brain. This is an important part of understanding how systems should be organized. It demonstrates why and how the function of the system dictates its structure. In particular, it explains the essential role of substructure, subdivision, and specialization by functional segregation in subdivisions."

We must have a balance of separation and integration, and the separate parts of the brain allow for that to happen. Bar-Yam uses the metaphor of a house. A house has separate rooms. For example, there's a living space, kitchen, bedrooms, and bathrooms. All of those spaces serve a function and that makes them more efficient in the overall habitat for a person. A house becomes functional because it has rooms.

10

The Science of How Nature Affects Our Brain

Can living in cities rather than rural settings affect our brains? According to science journalist and author Florence Williams, "Living in cities has been shown to differentially activate the amygdala." In Episode 056 The Science of How Nature Affects Our Brain, we talked with Williams about how our environment influences our health and well-being. She shared, "Our perception systems evolved in nature. . . . We should think of ourselves as full-sensory beings because that's what we are." Williams also shared research and stories from her book, *The Nature Fix: Why Nature Makes Us Happier, Healthier, and More Creative*.

We reflected on her idea that the secret to nature's power lies in its geometric fractal patterns. When we asked her to explain what fractals are and what new insights she learned about them, she said, "Fractal patterns are patterns that repeat at scale and they're often found in nature. So, for example, right now I'm looking out my window at some trees and there are little tiny branches and then they are sort of flowing into bigger branches and those are flowing into the tree trunk and the branches behind them are sort of in similar arrangements and similar angles and patterns. So this theory is that when we look at fractal patterns of a certain dimension . . . that makes our brains happy. It may even produce alpha waves in our brain and alphas are associated with calm and focus."

11

Unlearn Toxic Habits and Retrain Your Brain

How much does our environment and our biology affect our behavior? How can understanding our biofeedback empower us to live well? We explored these questions in Episode 058 Unlearn Toxic Habits and Retrain Your Brain with Eileen McDonald, co-founder and chief coaching officer of the Axeos Performance Institute. McDonald discussed how stress and poor breathing habits can influence our behavior and overall well-being. She quoted Alan Watkins, "There's no mystery in performance, our effectiveness and the results we achieve start with something deeper in the human system than behavior, and it's our physiology."

McDonald also shared how self-awareness, self-compassion, and self-care can help us unlearn harmful habits so we are free to practice more beneficial ones. "I think we are less forgiving of ourselves and more forgiving of others. And so in that process of new learning we're very critical and that critical mind comes out. And if we can just treat ourselves like we would treat a friend or a family member, somebody that we loved, and give [ourselves] that space to actually be able to have that time to learn and adjust and rewire . . . that critical mind actually holds us back a little because we're constantly saying, I should, I could, I would, and so we can change the language . . . and really open up and say, 'Wow, this was challenging but I was able to have one good hooray for me today.'"

12

Modern Life and the Female Brain

We talked with Louann Brizendine, a neuropsychiatrist and founder of the Women's and Teen Girls' Mood and Hormone Clinic, for Episode 059 A Closer Look at the Female Brain. Brizendine is the author of *The Female Brain*. We wanted to know more about how modern-day life affects the female brain because we were an office of mostly females, who lived the struggle to find balance in work, home, and personal life. With that in mind, we asked Brizendine how she thought modern-day life has influenced the female brain.

Brizendine explained that many years ago in Western culture we had layers of human connections with close family and friends who often shared the responsibilities of raising and educating children and caring for a home. That lifestyle worked in favor of the female brain and even created a biological safety net for females to manage their stress hormones. In contrast, Brizendine explained that our current lifestyle goes against the grain of female biology because it is driven by a cultural construct which promotes isolation. She said, "The cultural construct of work and life and educating and raising children is not tied to the land or to being able to live close by in close family compounds with your kin, your extended family. So you end up with the problems that we have today, which is really, I would say, against the grain of our biology. It's almost, for the female brain, like fingernails on a chalkboard."

We learned from Brizendine that many professional women will chose workplaces that support family life, and they will stay employed with these organizations longer. This explains why family-friendly policies, which allow for and encourage time off for caregiving, are becoming more critical to recruiting and retaining females in the workplace. Brizendine helped us feel empowered with a new understanding of our innate biology and the female brain as a complex system.

13

An Experience with the Institute of Noetic Sciences

The Institute of Noetic Sciences (IONS) offers a Worldview Exploration Program, which is "a research-based experiential program for middle school, high-school, and college students." The program helps young people become more aware of their perspectives and how they experience the world, so they can ultimately use self-awareness to become more open-minded and compassionate leaders. We were pleased to find out that the Worldview Exploration Program is just one of the amazing educational programs IONS provides for individuals from all stages of life.

Educational programs at IONS are deeply aligned with their mission to transform consciousness worldwide, a mission which stems from their founder, Edgar Mitchell. Mitchell was an Apollo 14 astronaut and the sixth man to walk on the moon. During his journey back from the moon he sat in the window seat of the spacecraft, and as the spacecraft would slowly spin around he looked out at the earth, the sun, the moon, and the stars. This experience guided him to a revelation that everyone and everything is connected within a universal consciousness, and this radically altered his worldview and inspired him to further explore the uncharted territory of the human mind and consciousness. He founded IONS in 1973 and his legacy lives on through the exploration of noetic science.

Noetic science is described by IONS as "a multidisciplinary field that brings objective scientific tools and techniques together with subjective inner knowing to study the full range of human experiences." Much of the work at IONS is research-based, and they also incorporate a broad range of human capability that goes beyond science, like imagination and intuition. IONS holds space for understanding that the relationship between the physical and the nonphysical is far more complex than we might have imagined. According to IONS, "The most powerful step people can take toward achieving their highest potential is the willingness and ability to understand new perspectives."

14

Finding Purpose Through Connection

In 2017, we attended The Science of What Connects Us conference, presented by the Institute of Noetic Sciences (IONS). We first learned about IONS from Doug, who had attended several conferences in the past and is a longtime champion of their work.

During the conference, we conducted several mini-interviews, one of which was with Andrea Dennis from Greenheart International. We shared her uplifting point of view in Episode 061 Finding Purpose Through Connection, "As individuals we move from our inner knowingness—from these noetic, direct experiences—and sometimes we move through life without ever having received any training, any deeper understanding, or guidance of how these inner knowing experiences can shape our life. So, I feel that if we, as a species, were to tap into that, there is no limit to how we could thrive and how we could work together."

15

We have this architecture of systems inside the body,
which represents information and gives us
our own internal intranet
of several hundred trillion microprocessors
that actually simplifies the complexity.

Rick Hanson
Episode 064 Our Evolving and Complex Human Brain

16

Exploring the Science of What Connects Us

In Episode 060 Exploring the Science of What Connects Us, Cassandra Vieten, president of the Institute of Noetic Science (IONS), explained some of the mind-expanding projects happening at IONS that are at the intersection of technology, innovation, and consciousness. She described their work this way: "Inner knowing is what we know about less scientifically, and that's what IONS is investigating. And I think the reason it's important is that when we rely only on the scientific way of knowing, that has brought us a lot of positive things, like cures for diseases and awesome technologies, but it's also brought us to a place where we're polluting the planet and we've created nuclear weapons and all of these things that could bring an end to us. . . . The idea here is to take both that powerful scientific way of knowing externally and the powerful way of inner knowing, and bring those together because that's probably where we're going to be able to innovate solutions."

17

*The most surprising thing is that paranormal activity is not paranormal, it's normal. . . .
Our minds are not confined to the inside of our heads, they are much more extensive than our brains. They're rooted in our brains but extend far beyond them, just like the electromagnetic fields of a cell phone are rooted in the cell phone, but they extend invisibly far beyond the material body of the cell phone.*

Rupert Sheldrake
Episode 062 Mysteries of the Extended Mind

18

Our Evolving and Complex Human Brain

In Episode 064 Our Evolving and Complex Human Brain, we spoke with psychologist and senior fellow at the Greater Good Science Center at University of California, Berkeley, Rick Hanson.

Hanson pointed out, "The brain appreciates complexity. . . . If there is no change of neural structure and function then, by definition, there's no learning, there's no development, there's no growth."

Hanson eloquently described how the human brain has evolved as a complex system and shared why it is so important "to really appreciate the implications of having a stone-age brain in the twenty-first century." He also warned, "This habituation to a high volume of information and stimulation is not good for the brain."

19

Our mental limitations prevent us
from recognizing our mental limitations. . . .
The sensation of being right or being certain
arises out of the same reward system
that has fueled opiate addiction,
cocaine addiction, alcoholism.
There's an enormous pleasure in being right
and an enormous discomfort
in being uncertain or ambivalent.
So, you have to actually force yourself to overcome
your own biology, which is really hard to do,
but it seems to be the only possible answer.

Robert Burton
Episode 063 A Philosophical Look at the Mind

20

Mitigating Risk in Biotechnology

While we were podcasting for the International Conference on Complex Systems 2018, we had the opportunity to interview George Church for Episode 099 Mitigating Risk in Biotechnology. Church is a professor of genetics at Harvard Medical School and professor of health science and technology at the Massachusetts Institute of Technology. Church talked to us about the evolution and risks of modern biotechnology. He said, "Rather than taking millions of years to evolve a new species, we can have some of the capabilities of a new species in weeks or months, and it can spread throughout the population very rapidly as well."

Church also discussed the goals and mission of The BRAIN Initiative, which he co-initiated in 2011, and why he believes people should have access to their own genome. He said, "We're getting excited about the progress we've made in making any organism resistant to all viruses by changing the genetic code. We've already made that whole genome scale change in one industry microorganism and now we're trying to apply it to a variety of other microbe plants, animals, and human cells."

21

Mysteries of the Extended Mind

We talked with Rupert Sheldrake for Episode 062 Mysteries of the Extended Mind. Sheldrake is a biologist, author, and researcher in the field of parapsychology. His most recent book, *Science Set Free*, exposes the dogmatic nature of modern science and questions the way many scientific assumptions are impacting the future of humanity. Sheldrake backs his ideas up with history, philosophy, and scientific evidence. What we love most about Sheldrake is that he questions everything, even his own assumptions. He is a complexity thinker and has a curious mind through and through.

Sheldrake talked with us about theories of human consciousness, paranormal phenomenon, mystical experiences, and spirituality. He shared how he began his career as a biologist studying plants and animals and described how he's deeply involved in research and discussions about human consciousness. When we asked him about his fascination with consciousness, paranormal activity, and other unexplained powers of the human mind, he described a natural progression of curiosity throughout his life. He first studied homing pigeons, then dogs with the ability to detect when their owners were coming home, and that research lead him to study "paranormal" human capabilities like telepathy.

We asked Sheldrake how his work is changing the world, and he responded by advocating for the theory that our minds are more extensive than our brains. He stated that this idea of an extended mind "enables us to escape from the prison of the brain, the idea that the mind is imprisoned inside our skull, which is a very isolating idea. It disconnects us from each other and from the world around us, and it's also false." He addressed the extended mind theory in *Science Set Free*, in which he shared a number of stories about telepathy, premonition, and the sense of being stared at, all of which provide evidence that our minds can extend beyond our brains.

22

Coherence and the Heart's Intuitive Guidance

We talked with the executive vice president and director of research at HeartMath Institute, Rollin McCraty, for Episode 073 Coherence and the Heart's Intuitive Guidance. McCraty shared stories and examples from scientific research about the complex nature of heart coherence, intuition, and consciousness. He said, "As we learn to follow intuitive flow and really listen to intuitive messages, our inner guidance, that is literally what elevates consciousness and raises the vibration of our consciousness state."

McCraty also talked about HeartMath's Global Coherence Research, which encompasses scientific data on the interconnectivity of all living systems through Earth's magnetic field. McCraty said, "Our thoughts, our emotions, our intuitions are all a part of the emergent properties of a complex system such as us living human beings. And what we've shown in our [HeartMath's] early work is that when we feel things like frustration, impatience, and anger, especially, that literally creates measurable incoherence within the body's systems."

McCraty spoke about the the importance of practicing self-awareness and self-regulation in order to bring ourselves into balance with the heart's intuitive guidance and reach higher levels of consciousness. He said, "What a lot of people don't realize is that the heart has its own complex nervous system, dubbed 'the heart brain,' and this is meant literally, not metaphorically."

Hardwiring Happiness

Not only did we interview best-selling author and psychologist Rick Hanson for Episode 065 Hardwiring Happiness, we also dove into his book of the same title. This *New York Times* bestseller is about "the hidden power of everyday positive experiences to change your brain—and therefore your life—for the better."

Hanson explained the human brain as a complex system and specifically addressed two different modes which ultimately determine how we deal with everyday experiences. He described these two modes as the reactive mode and the responsive mode. The reactive mode is a brain setting that evolved to keep us alive. It is our alarm system that serves us in the short term, but is not at all concerned with our long-term needs, including our growth, fulfillment, or happiness. This brain setting is very different from the responsive mode, which is far more beneficial to our growth—and ultimately our happiness—because it allows us to refuel and repair. Hanson called the responsive mode "the green zone" where we feel safe, satisfied, and connected. It is our "home base" where we contain and calm our reactive states as they occur.

We discovered a fascinating connection between Hanson's description of brain modes and complexity. We realized the reactive mode is engaged in reductive thinking, while the responsive mode emphasizes complexity thinking or holistic thinking. When we are in the reactive mode, we think reductively, our perceptual world narrows, and we tend to lock onto one problem that is causing us pain and fixate on it. We also view the world with fear, focusing on scarcity and exclusion. On the contrary, when we are in the responsive mode, we are in a state of positive well-being, so even if our mind is experiencing pain or upset, our perceptual field broadens and we are much more likely to perceive and act in complex ways. We tend to view the world feeling protected and connected and we focus on abundance, inclusion, and interconnection.

It is challenging to pull ourselves out of the default reactionary mode—and it is also more important than ever that we try. Hanson said, "The brain is the most influential organ in the body," and we can use self-directed neuroplasticity to build resilience. We can influence our brain with daily meditative practice to purposefully "take in the good" in our lives. By regularly taking in the good, we are able to draw ourselves out of reactive episodes, while also building responsive capacities in our brain.

As we build resilience into our neural structure, we are able to maintain an inner sense of peace, contentment, and love that is less likely to be jarred by external forces. Eventually, with enough practice, we can rest in the happiness that emerges from within our being. We just have to come back home to our responsive mode, where we thrive in abundance, inclusion, and interconnection.

24

Take In the Good

Rick Hanson talked about "taking in the good," which is "the deliberate internalization of positive experiences in implicit memory." We can achieve this through Hanson's simple four-step model, which spells the acronym HEAL:

Have a positive experience. Think of a positive experience and write about it or draw an image of it.

Enrich it by adding details to the experience.

Absorb it by spending a few minutes thinking about and absorbing your enriched positive experience.

Link positive and negative material by remembering your enriched positive experience when you experience something negative.

Hanson claimed that we can use these steps to "HEAL" ourselves by letting good experiences soak into our being. He does not advocate denying or resisting the bad in our lives, because as he described it, the truth is what sets us free and the truth includes everything, both the good and the bad. The HEAL method works to build resilience by slowly chipping away at our built-in negativity bias. It hard-wires positive thoughts and feelings in our brain so they become positive neural traits.

25

The Science, History, and Implications of Real Magic

What is "real" magic? How is magic tied to consciousness? How will magic play a role in the future of science? We interviewed author and chief scientist at the Institute of Noetic Sciences, Dean Radin, for Episode 081 The Science, History, and Implications of Real Magic. Radin discussed his scientific research on mind-matter interaction, consciousness, psychic phenomenon, and esoteric magic. He said, "Magic is not only linked to consciousness, magic is completely dependent on our idea of what we think consciousness is."

Radin also shared details from his new book, *Real Magic*, including what scientifically-informed magic actually is and how magic will play a key role in exploring the frontiers of science. He said, "How do we create an expansion to our existing scientific worldview that would accommodate these phenomena—telepathy, clairvoyance, precognition and psychokinesis? I started to work on that problem and after a while I suddenly realized that there were clues about how to do this and the clues all arose from the esoteric traditions, and that immediately led to the traditional notion of magic."

26

Leveraging Self-Directed Neuroplasticity

How can we leverage self-directed neuroplasticity so we can grow and change as much as possible? Rick Hanson explained how understanding evolutionary neuropsychology and using a daily meditative practice can help us feel safe, satisfied, and connected. Hanson said, "Everyone has the power inside themselves every day to grow a little bit, to heal a little bit, to become a little wiser, a little stronger, a little happier, a little more loving. Everybody has that capacity to help their brain and their body all together change for the better every day."

In Episode 064 Our Evolving and Complex Human Brain, Hanson discussed how essential complexity is for understanding human potential and happiness. He said, "Nothing really makes sense unless we look at it within the framework of complexity," and he summed up his advice this way, "Think complexly, act simply."

27

Coherence means the parts of the system are somehow communicating with each other and working together harmoniously. So, there is a lot implied there, that the parts of the system come together in a coherent way to give an output or a function that is greater than the sum of the parts.

Rollin McCraty
Episode 073 Coherence and the Heart's Intuitive Guidance

28

Evolutionary Theory and Behavioral Science

We shared our conversation with evolutionary behavioral scientist Gad Saad at the International Conference on Complex Systems 2018 in Episode 103 Evolutionary Theory and Behavioral Science. Saad talked about how elements of complexity theory apply to his research as a behavioral scientist, and in particular how his efforts are about *understanding* complex behavior; not justifying behavior, or trying to "fix" it. "As a professor of behavioral science, I don't care about how much work is ultimately applied or misapplied. All I try to do is understand the human condition." A key to this understanding is looking closely at adaptations. "For something to be an adaptive mechanism it doesn't have to be conscious. Breathing is an adaptation."

29

Systems-Based Precision Medicine

Episode 104 Systems-Based Precision Medicine captured our conversation with Daniel Stickler at the International Conference on Complex Systems 2018. Stickler is a physician and co-founder and chief medical officer of the Apeiron Center for Human Potential. Stickler explained Apeiron's systems-based model, which focuses on human systems, homeostatic balance, and optimization. He said, "Optimized care is about how do we . . . create a system that functions at the level that it's supposed to." He also discussed how and why lifestyle decisions are the key to creating an optimized life. "Where health care really needs to focus its attention is on coaching."

30

Our current understanding of systems,
which is coming from neuroscience and biology,
is taking us to a new level of awareness.

Angela Montgomery
Episode 036 Harnessing System Potential

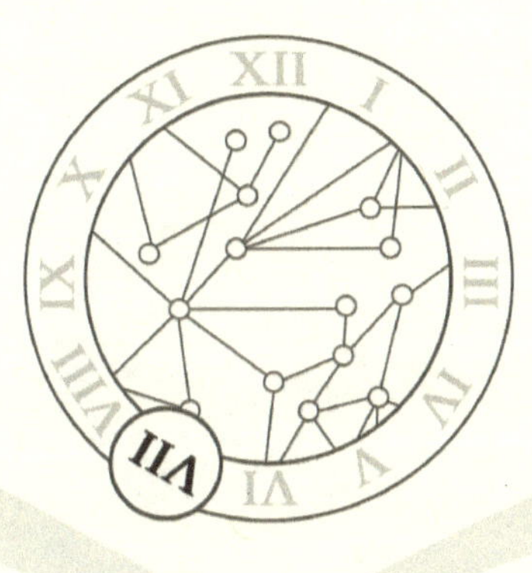

JULY

1

Nothing is Impossible with Creativity

Doug taught us that creativity is a key element of effectiveness, and that there is no such thing as impossible—only a lack of applied creativity. Over the years, Doug shared various creative tools and methods with us. One method was: for any given issue or situation, think of at least ten different ways it can be addressed, one of which should be "zany" or "out of this world," as Doug would say. This quick creativity method would help us stay expansive with our ideas and avoid getting too narrowly focused. This method also helped us avoid slipping into a reductionist mindset.

Doug taught us about synectics, which is similar to springboarding or brainstorming. The synectics technique promotes creative thinking with a small group of people to generate new ideas or solve problems. Other creative tools that Doug taught us included permutations, combinations, forward and reverse chaining, recursive excursions, and transference. While Doug taught us many creative tools, our biggest learning was understanding that creativity is our greatest asset and resource. Our creativity was a key element to *Human Current*'s success.

2

Design Thinking Basics for Navigating Complexity

How do we take an understanding of complexity and apply it to how we handle wicked problems? Is design thinking an innovative action plan for approaching complex problems? In Episode 027 Applying Design Thinking to Complex, Wicked Problems, we talked with our former co-host, Stacy Hale, to address these questions and take a closer look at what "design thinking" really means.

Design thinking is not just for people who love sticky notes. It is not a one-size-fits-all process, either. Design thinking is a way of seeing, and it is useful for navigating complex or unpredictable situations. A key to design thinking is holding multiple perspectives. Consider the experience customers might have with the product or service you're creating. You might take what data you have—for example, they like dogs, she likes extreme sports, he likes ciders—and with a pen and some paper connect the dots by imagining them moving from one experience to another. You could ask questions like: What would a typical day look like? How, when, and where might they interact with your offering? At their homes? In their social circles? You could even map these questions and responses out as a network.

As we understand it, the term "design thinking" was coined by IDEO founders David Kelley and Tim Brown, and in its simplest form, the process goes like this:

Ideate. Explore and define the problem.

Prototype. Design, develop, and model solutions.

Evaluate. Test and incorporate feedback.

We think people have been design thinking for a lot longer than there's been a name for it, because it's a natural way of exploring by asking questions and adapting mental models when more information is gathered. Sometimes when a solution works it becomes a best practice and solidifies into strategy. However, design thinkers never lose sight of the importance of context or complexity—or of feedback. They know that every situation is unique. It's this sort of flexibility that makes designers sought after for a whole range of business challenges, often outpacing classic strategy where innovation is concerned. The good news is, anyone can apply the design thinking process.

3

*The world is complex whether you like it or not,
and we need to unpack that and
understand that we're saying more than that,
there is a theory and understanding of complexity
that we can bring to bear, and with that
we do a better job of engaging with
our human and more-than-human world.*

Jean Boulton
Episode 075 Embracing Complexity

4

Leaping into Design Thinking

In Episode 028 Leaping into Design Thinking, we spoke with Turi McKinley from frog design, a global design and strategy firm. As a leader at frog, McKinley regularly utilizes design thinking principles in order to accelerate thought leadership initiatives. She said, "You need to have multiple perspectives together in order to make a solution that is meaningful to the user, appropriate for the business, and sustainable in the market." McKinley explained the complex topic of design thinking in layman's terms, as well as why design thinking is an essential mindset for anyone dealing with complex, wicked problems. Her bottom line: "Stop talking and start making!"

5

Design-Led Innovation

"Ideas are free. If you don't make them, they are up for grabs," said Jon Kolko when we interviewed him for Episode 029 Design Led Innovation. We agree! Co-host Angie Cross replied, "What good is an idea if it stays in your head?"

Kolko is the vice president of design at Blackboard and founder of the Austin Center for Design. Kolko described design as "about making things—it's a verb, not necessarily a noun." Design is also "about making things for people and thinking about them critically. It is both strategic and tactical." We discussed this quote from Kolko's book *The Strange Connection between Entitlement, Social Innovation, and Interaction Design*: "I have no misconceptions that designers can 'solve' massive problems, or even approach them on their own without collaboration from other disciplines. But I feel strongly that designers make great agents of change and can champion new and novel approaches to old and tired problems. The best indicator of design success, in my experience, is a passion to make an impact, and I see a generation that is wildly passionate about addressing social problems."

6

How to Apply Human-Centered Design

In Episode 030 How to Apply Human-Centered Design, we talked with Andrea Ippolito, lead at the Department of Veteran Affairs (VA) Innovators Network. Ippolito shared how human-centered design is implemented and deployed at the VA in order to create exceptional experiences for our veterans. She said, "Using design thinking actually helps reduce rework, because you are so focused on your users. You're impacting the bottom line on both ends of the spectrum, both with operation and sales." She also shared, "Design thinking forces you to understand the context in which your user operates and lives, and what they appreciate and don't appreciate." Ippolito shared examples and advice that are helpful for anyone wanting to build an innovation strategy using design and systems thinking methods. She shared a quote by Tim Brown, co-founder of IDEO: "It's not 'us versus them,' or even 'us on behalf of them.' For a design thinker, it has to be 'us with them.'"

7

We are all holding a piece of the puzzle,
and the only way we can get to the big picture
is to see how the pieces fit together.

Stacy Hale
Episode 027 Applying Design Thinking to Complex, Wicked Problems

8

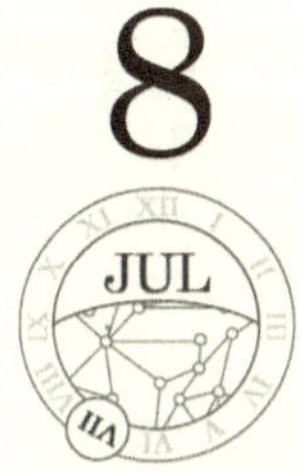

Harnessing System Potential

We talked with author Angela Montgomery for Episode 036 Harnessing System Potential. Montgomery discussed the importance of leadership education and systemic thinking within organizations, and shared some key insights and tools for managing business complexity. She said, "Today we need a much more sophisticated type of organization that gives space for humans to be humans inside a human system." She illustrated this with a quote by Betty Bender, "When people go to work, they shouldn't have to leave their hearts at home."

9 JUL

Preparing Leaders for Change Management

As we explored the complexity of leadership, we couldn't help but think about change management. Change management has various applications, including adapting to change, controlling change, and effecting change, all of which require effective leadership. As any effective leader will admit, successful leadership—related to change management or not—begins with oneself.

Oddly enough, the steps to change management tend to vary from expert to expert and they do not take one's leadership attributes into account. For example, in John Kotter's *The 8-Step Process to Accelerate Change* e-book, he describes a process for leading organizational change: increase urgency, build the guiding team, get the vision right, communicate for buy-in, empower action, create short-term wins, do not let up, and make change stick.

While these steps are useful, there is little emphasis on the leadership required for change. The same could be said for George Vukotich's approach from his book, *10 Steps to Successful Change Management*, and Mark Murphy's article in *Forbes*, "3 Stages of Successful Change Management," and other standard, stepwise change management procedural guides and certifications. These experts offer valuable processes which are worth exploring. However, we shouldn't undervalue the importance of first starting with ourselves as leaders of change.

We recommend beginning with self-assessment and personal change. Here are three practices to help with this:

Get clear about your role. You'll be able to communicate more effectively with others if you clearly envision and understand your role and what you can offer throughout the change management process, including your expectations of yourself and others. Keep in mind, by getting clear with your role, you may also come to recognize that you are not the best person to lead change management, and this is OK. After all, if you are not the right person to lead change management then the change will ultimately not be successful. Your awareness of this makes you a great leader.

Explore and understand your values. Our values play a crucial role in what we do in life, how we react, and our priorities. Not everyone has the same values, and organizations also have their own values. Knowing your own values can help you understand your reactions and feelings, ultimately leading you to make effective decisions and take appropriate action. Understanding other people's values will help you navigate relationships and priorities. For a quick and effective tool, check out the free value assessments that Barrett Values Centre offers at valuescentre.com.

Choose the best approach. There are many change management processes that you could choose to follow, some of which we have mentioned. Research the various models to truly understand the process and outcomes they are suggesting and then choose one, or select a combination of models that will ultimately empower others. It is also important to not just settle for a model that currently fits with your organization, but rather choose one that aligns with where you want the organization to be.

10

Self-organization is a process question for me fundamentally. It's about, what are the gates that allow you to create teams? What are the gates that allow information to flow back to the rest of the organization? And when does that team know that they can disband?

Bud Caddell
Episode 069 Designing Meaningful Work in Complex Times

11

The Complexity of Leadership

In Episode 046 A Sneak Peek at Our Leadership Series, we discussed the complexity of leadership and shared clips from upcoming conversations with Scott Jancy and Paolo Gaudiano. We asked Jancy and Gaudiano how they became interested in complexity. Gaudiano said, "I realized that the brain was the most amazing, complex, and fascinating machine that existed on earth."

Jancy resonded: "I view every organization from the perspective of a system in balance. And when you approach an organization that is struggling with some problem—from that perspective, the holistic perspective—and understand that, yes, you have a problem to solve, but more importantly you need to create a system that is balanced, the solutions sometimes emerge on their own with a little bit of coaxing."

12

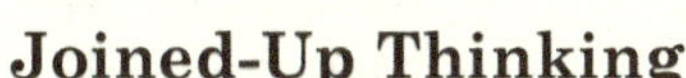

Joined-Up Thinking

It hasn't been that long since organizational science experts looked to the field of engineering for guidance in managing change. Organizations were studied like machines that could be optimized by improving the performance of the components—people.

In reality, organizations are more like biological systems, with complex interdependencies that must be understood to effectively navigate change and achieve collective goals. If you want to truly understand the behavior of an organization—a human system—you have to first understand that it's not the performance of the individual parts that matters the most. In a system where there are interdependencies (like an organization), optimizing the parts separately will not necessarily improve the performance of the whole. What you want to look for is how the relationships between parts lead to patterns that arise across the network.

Seeing and recognizing the nature of these links allows people like our guest for Episode 036 Harnessing System Potential, Angela Montgomery, to model and design change scenarios that help organizations adapt and evolve toward a unified goal. Montgomery co-founded Intelligent Management with partners Domenico Lepore and Giovanni Siepe. Their firm supports those who seek systemic change in their operations to manage complexity and continuously improve.

Montgomery explained that many business schools and universities are organized in a siloed, hierarchical manner and continue to perpetuate a model of organizations that appear to be more like the inputs and outputs of nineteenth century factories than a dynamic human system. Leaders and managers need new tools to use with the challenges they're coming up against.

Being able to push past our mental models and employ a more "joined-up" way of thinking is the first step to getting to the bottom of what's really going on. This method requires designing workplaces that overcome the problems of being artificially divided into silos and organized into traditional hierarchies, which don't allow us to perform at our full collective human potential.

In our global society, political and ideological polarization have emerged from systemic inequality and continue to create systemic effects that build to the level of crisis, stunting the growth of information and stifling cooperation and innovation. Helping people, especially leaders, to see what's really going on is going to require more than theories because there is no universally accepted method for navigating the space of unintended consequences that complexity produces. It's a design challenge, and we have to prototype the solutions.

13

I think that flow states are the key
to living within complexity
and maintaining an awareness of it,
because it exists naturally in so many situations.

Stacy Hale
Episode 004 Complexity and Flow

14

The Complexity of Living Organizations

We talked with Oleg Konovalov, author of *Organisational Anatomy*, which views organizational processes and functions from a biological perspective. In Episode 037 The Complexity of Living Organizations, Konovalov said, "Complexity is always multidimensional, it is not flat." He shared key concepts and metaphoric examples to help unpack complexity and make it more digestible, consistent with his principle that, "Professionalism is reflected in simple language. . . . Complexity is a pack of simple things; don't make it unsolvable, don't make it a mess."

15

Making Sense of Human Systems

The following is an excerpt from Episode 044 Making Sense of Human Systems. In this episode, we interviewed founder and chief scientific officer of Cognitive Edge, Dave Snowden. Let's eavesdrop on part of their conversation:

***Human Current*:** Professor Snowden, what is a takeaway related to complexity? Maybe something that our listeners can do?

Snowden: You manage in the present by parallel "safe-to-fail" experimentation, you don't manage to future states. That's the most key thing. So, for example, you don't announce the change initiative because everybody has been through too many of those and you'll get a negative reaction. You start to make changes, but you don't do *fail-safe* design. You do *safe-to-fail* parallel experimentation, and that's probably the key thing.

***Human Current*:** Ah, nice and simple. It *sounds* simple, doesn't it?

Snowden: It's simple but it's not simplistic. The trouble is, a lot of people, they subordinate complexity language into old methods. I'm not sure whether that's putting old wine into new wineskin or new wine into old wineskins, but either way it's a bad idea. And they don't realize just how radically different complexity is. You manage the present and you manage the things you can manage, and you don't manage to define future states. And that really is radical.

16

Structuring and Amplifying Change in Public Service

by Scott Jancy for *Human Current*

Structuring change is easy when looking from the outside in. Architects, urban planners, concerned citizens, and politicians, among others, can find themselves speculating endlessly about what needs to be done to change a system.

Addressing the concerns of current government workers in the system through training and open communication is a positive step, but the real change will come with an influx of new people into the government ecosystem. The new attitudes, new ways of working, and new connections will open people's eyes to what could be. Changing narratives starts with seeing new perspectives.

The more that people can see the big picture and connect the dots for how change is going to take place, the easier it will be for it to actually happen. Anticipating what comes next is easy when people involved are generally on the same page and moving in the same direction.

The decentralization of decision-making systems is a key aspect of creating faster, more relevant change for people. Give people space to solve their own problems and they will often find solutions that are more effective and useful to the people actually using them. This idea further reinforces the idea that one needs to work with the resources around them in order to create homegrown solutions to local problems.

The unfortunate reality with creating change in organizations is that when the door of opportunity opens, you need to know when to step in or be at risk of being left behind. This will be a challenging time for leaders, constituents, and employees alike. If you are keeping an eye on how the change is occurring or maybe you are resistant to it, you may miss your opportunity to be a part of a movement to make a more effective and useful form of government that provides more value and service to people.

This system is only as effective as people want it to be and the real value will be felt far away from centers of government. Understanding what people need is the key to understanding how to advocate for them.

Advocating for citizens and creating change in government can happen from the bottom up or the top down. However, engaged citizens and enlightened leaders working together to co-create the future is the secret to effecting lasting change. Digital tools can serve as the medium to help bring both groups together to create that path forward. Getting people to care is at the heart of civic engagement.

The most successful change efforts start at the grassroots level. Enabling people to see a problem, define it, and then put together a plan to solve it is the best way to achieve long-term success. Top-down change that is mandated from above can make people feel even more disconnected, regardless of how well-intentioned one's efforts may be.

Understanding what people truly need starts with creating mutual respect. Otherwise, your engagement efforts will be superficial at best. When people see that others are listening and that your efforts are sincere, then they will be more inclined to care about what you are doing—and may just join you to help out.

Excerpted from a blog previously published at human-current.com/blog on December 5, 2016.

17

As the interconnectedness of society increases,
we have to create systems that can handle that complexity.

Stacy Hale
Episode 050 Complexity Science Is Everyone's Science

18

Embracing Complexity

We spoke with Jean Boulton, author of *Embracing Complexity: Strategic Perspectives for an Age of Turbulence*. In Episode 075 Embracing Complexity, Boulton said, "Complexity is, in some sense, where physics explains evolution. It's the science of open systems." She talked with us about different concepts, including: how complexity thinking compares to systems thinking, change management, organizational strategy, complexity as a worldview, and even how this field is shining a light on climate change. She said, "We live in a dynamic, evolutionary world of *becoming* rather than a stable, predictable world of *being*. . . . We need another paradigm that informs us to make better decisions." She reminded us of Heraclitus's saying, "No man ever steps in the same river twice."

19

If we can help an individual or a group
understand their own part and influence in the system,
then we can connect them to not only their own environment,
but to the bigger and bolder ecosystems
that they are a part of, and this is
where they become complexity thinkers.

Denise Easton
Episode 117 Influencing Pattern-Based Change

20

The Complexity of Change Management

We were familiar with the book, *Organizational Communication Approaches and Processes* by Katherine Miller, in which she described how some twentieth-century organizational theorists are no longer effective in our fast-paced, interconnected, and global society. In response to linear management models becoming ineffective and irrelevant, organizations have been increasingly applying theories of chaos, self-organizing systems, and complexity. Furthermore, as any complexity thinker and leader will tell you, change does not happen in a vacuum or overnight; instead, change—especially as it relates to organizations—is continuous and complex.

While change management literature tends to be prescriptive and results-focused, we believe that coupling it with complexity theory can help increase our understanding of the continuous nature of change and nonlinearity in today's organizations. The benefits of applying complexity thinking to organizations—which are complex adaptive systems—can lead to seeing situations from different lenses, developing new and innovative approaches to problems, and gaining a greater appreciation for processes such as safe-to-fail. Many organizations are even going as far as to create a safe-to-fail environment to manage their organization's continuous changes. Additionally, while various aspects of change management assert a level of control (adapting to change, controlling change, and effecting change), complexity theory embraces the impossibility of control in complex adaptive systems. As Jurgen Appelo told us in Episode 068 Managing for Happiness with Complex System Science, "With a complex system, there is no such thing as central control."

In Episode 048 The Intersection of Diversity, Innovation, and Performance, we shared our interview with Paolo Gaudiano. Gaudiano noted that larger organizations are more complex than smaller ones since they have more people and more constraints. However, there is higher variability in the likely outcomes with smaller organizations. He explained that even though there is typically less complexity within smaller organizations, they are faced with a greater cone of uncertainty, which broadens the probability of outcomes, leading to limited predictability.

This realization led us to believe that smaller organizations—rather than larger organizations—are more likely to reap the results of a standard change management approach. They are more likely to experience complicated problems versus complex problems. Larger organizations, on the other hand, should consider coupling a change management approach with complexity, chaos, or self-organizing systems theory.

21

One of our fundamental flaws is that
we believe evolution is a synonym for progress,
and it's not.

Meg Wheatley
Episode 090 Reclaiming Leadership for the Human Spirit

The Challenges and Rewards of Change

As described by the author William Bridges, change is something that you can mark on a calendar—for example, the day you met your life partner, started a new job, or a loved one died. Change triggers an emotional transition, which is experienced in our head, heart, and gut. "Transition is the inner psychological process that people go through as they internalize and come to terms with the new situation that the change brings about," said Bridges. It's the transition that often makes us feel emotional, brings on feelings of vulnerability, and can be hard to comprehend. Successful change actually relies on our ability to get through this emotional transition.

Bridges discussed his theories of change and transitions in his book, *Managing Transitions: Making the Most of Change*. The emotional transition that comes from change can be both challenging and rewarding, and everything in between. Bridges revealed a three-phase process for moving through emotional transition. These phases include: ending and letting go, the neutral zone, and the new beginning. The neutral zone, which usually does not feel neutral at all, is where we can experience our biggest ah-ha moments, tap into creativity, foster emotional intelligence, and be expansive. However, the neutral zone can also present struggles with feelings of vulnerability, fear, and uncertainty.

Research professor and author Brené Brown discussed vulnerability in one of her *TED Talks*, "The Power of Vulnerability." She encouraged us to embrace vulnerability, "lean into the discomfort," and also into the joy. Brown stated that we are hard-wired for struggle as children and we should remember this as adults. From Bridges' description of the neutral zone, we assume that that is the place where Brown cautions us against our urge to "make the uncertain, certain."

Along with what we've learned and applied from Bridges and Brown, we have also put into practice the disciplines of Peter Senge, a systems thinker and author. Senge's book, *The Fifth Discipline*, focused on shared vision, mental models, personal mastery, team learning, and systems thinking—all of which are helpful when managing change and transition. Specifically related to change, Senge claimed that "leadership is always about change," and that "change naturally induces fear in all of us: fear of the unknown, of failure, of not being needed in a new order of things." Senge reminded us about the importance of not getting stuck in the change itself, but to also focus on what can be conserved. Additionally, Senge shared the importance of shifting deep-seated mental models—assumptions, beliefs, and mindsets—that can get in our way of successful change, another good reminder as we manage change and transition.

23

People think that change is simply
doing something different,
but it isn't always that.
Change is often shifting
and influencing what's happening
at many different levels.

Denise Easton
Episode 117 Influencing Pattern-Based Change

The Complexity Worldview

In Episode 075 Embracing Complexity, we talked with Jean Boulton, the co-author of *Embracing Complexity: Strategic Perspectives for an Age of Turbulence*. Our conversation with her covered concepts from how complexity thinking compares to systems thinking, change management, complexity as a worldview, and even how this field is shining a light on climate change. We realized that she lives and breathes complexity, using this worldview to frame how she thinks, feels, and acts. We wanted to learn how she became interested in this field of complexity theory.

Boulton shared that her interest in complexity theory was the result of path dependency, calling it a "systemic coming together" of her outlook on life, her background as a physicist, and her concern for social justice. Her interest is lifelong; when growing up her friends would describe her as a "serious seeker" and questioner. She talked with us about her passion for challenging the assumption that the world is mechanical, predictable, and controllable, a worldview that is often implied in modern science.

Boulton suggested that organizations can do a better job by leveraging some of the ideas of complexity theory. She mentioned detail planning for the short term, experimenting often, and getting more cross-functional groups involved in feedback—all aspects of design thinking. She also spoke about honoring emergence as a way to help manage change and even promote innovation. Boulton recommended taking time to pause, review, and ask questions like: What has emerged that we couldn't have known or planned for? Do we want to shift our focus a little because of what has emerged? Does it align with our values and bigger vision? These questions and insights are wonderful ways to probe an organizational system and explore the wider picture before reacting in a way that could result in unintended consequences. Embracing emergence is also critical to an organization's survival because this practice recognizes unknowns and allows the system to adapt accordingly.

Many of Boulton's insights are relevant to personal change as well as organizational change. It is becoming more and more apparent that you can't have one without the other. We are often reminded of the importance of self-awareness while working within complex systems, and her advice was a helpful reminder that we need to continually question our own assumptions and mental models, especially while working toward personal growth and transformation.

Boulton strongly believes, as we do, that the complexity worldview can help us navigate our world as it is, not as we believe it to be or want it to be. Practicing humility and curiosity helps us on our journey because we can never know everything, but we can learn enough to gain some clarity and perspective. Boulton explained that complexity "is a middle ground theory between saying we know everything and we know nothing." It's about learning to be comfortable with uncertainty, because inevitably things will not go according to our plan. We can adapt by becoming more resilient and refraining from our command-and-control methods.

Ultimately, Boulton taught us that embracing our complex reality is simpler than trying to control a machine that does not exist. If we begin to understand complexity and unpack it, we can start our journey toward figuring out how to act congruently with and within it.

25

You cannot understand the complexity of the system without mapping the network out; on the other hand, network science has learned a lot from complexity science on how to analyze these networks and how to look for organizing principles within them. I think only together will we make progress in understanding complex systems.

Albert-László Barabási
Episode 115 The Network Science of Success

26

Conversations Are Power

We read a great post on the Systems Thinking World discussion group on *LinkedIn* by Andy Walkden, "Are Systems Thinkers the Cassandras of the Modern World?" We've wondered ourselves if, like Cassandra from ancient mythology, systems thinkers are doomed to never being believed, not so much because there aren't yet enough systems thinkers, but more because the nuances and behaviors of complex systems are not yet woven into our language.

People tend to avoid thinking in terms of complexity because they think it means complicatedness. After all, how deep do you have to go when you are just looking for a cause and a solution to a problem? We will not think to look for systems-based solutions until we are used to talking about systems in a way that is not overly technical, and until we have tools and competencies that enable complexity awareness in everyday discourse. Complexity is a fact, uncertainty is inherent; however, we still need meaning, and we still need things to make sense.

To facilitate a widespread understanding of how important systems thinking is, we have to use our technological muscle as a megaphone to incite a large-scale cultural shift, one that moves away from blame and singular cause-and-effect reductionism, and instead begins to tell stories about interrelationships. Relationships have greater causal agency on a system than the actions of one entity or organization. Until our media is incentivized to serve up something other than blame toward one node in the system, we're doing little better than holding trials in a gladiator arena. So we have to ask for something different with the agency and influence we can generate together. New attractors have to form. Conversations are power.

It's stories and relationships that have the real agency in our social and even economic dynamics. The activity they generate—on social media for example—is what forms attractors. So if all the Cassandras gather and have public discourse around some of these topics, guess what? In complex systems-speak: preferential attachment leads to attractors. In plain English: change happens.

27

Self-Reflection Can Strengthen Complex Systems

Reflection is the act of purposeful learning. With practice we can get better at learning from experiences in our work and our lives. This requires us to be curious and inquisitive about our actions, as well as the feedback we receive from others.

Looking back allows us to identify and expand upon what really matters, because we receive ideas on how to move forward in a way that better serves our greater purpose. And, as acting agents within complex adaptive systems, we need to recognize that communication, feedback, and reflection are all crucial to keeping systems alive. Feedback shapes how change happens. Together, we can practice the art of reflection and find deeper connection through exploring the past.

Here are some curiosity questions we can ask ourselves: How do we spend our time and money? What and who influences us? A customer, a department, a book, an advisor, a competitor, the economy, worry, fear, creativity, inspiration? Are we in alignment with what really matters?

By seeking involvement and advice from others while approaching this reflective process, we can identify assumptions and conclusions with less bias. Collaboration helps shield us from the lens of our mental models. Co-workers, family members, and close friends ultimately strengthen our reflective processes by providing feedback, which can help us identify patterns of behavior.

Our human systems are constantly evolving. We can learn from change, better predict likely changes in the future, and prepare for them accordingly. However, this is only possible when we take the time to reflect, learn, and recognize patterns. So, challenge yourself and your networks to engage in reflection, because everyone—including the system itself—will be stronger for it.

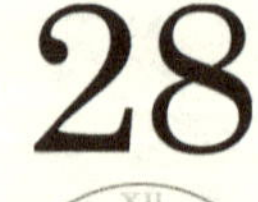

28

The question of whether you are really successful in research is vastly more about, did you ask the right question? Rather than, were you good at the mechanics of answering the question. Strategy is critical in research.

Stephen Wolfram
Episode 101 A Journey of Computational Complexity

29

Design is the Answer

Russell Ackoff was an architect, systems thinker, and philosopher who constantly studied human systems and human behavior. One of the interesting systems of human behavior Ackoff examined was problem-solving. His overall mantra seemed to echo the famous saying often attributed to Albert Einstein: "We cannot solve our problems with the same level of thinking that created them."

Like Einstein, Ackoff thrived on creative thinking and would constantly shift boundaries to reframe problems, including the "problem of problems," and problem-solving. How very meta of him! We watched Ackoff's "Systems Thinking Speech" on *YouTube* where he explained his thinking when he said, "A problem is to reality what an atom is to a table. People experience tables, not atoms." We experience the whole, and reality is a whole mess of problems interacting simultaneously. "Reality is a system of problems," said Ackoff.

So, if problems are just interacting concepts that create reality, the real question we should be asking ourselves is not, how can we solve this problem? Rather, how can we change our reality?

Sounds pretty intimidating, doesn't it? Think of it this way. Ackoff described four different ways people can treat problems: absolution, problem resolution, problem-solving, and dissolving a problem.

Absolution is the way we treat most problems. We ignore it and hope it will eventually go away.

Problem resolution means we look into the past to find a resolution which is "good enough."

Problem-solving is what we were taught in school. We were told this is the best thing you can do with a problem. We look for the "optimal solution," the best thing to do in the current situation. However, no problem ever stays solved in a dynamic environment and every solution creates new problems, so it's a temporary and ineffective way to treat a problem.

Dissolving a problem is the best way to create lasting change. We can only dissolve a problem through design, by redesigning the system that has the problem, so that the problem no longer exist.

As an architect, Ackoff received professional training in design and obtained a profound understanding of how systems, problems, and design are beautifully connected. He explained that an architect designs the entire house first, and only changes the design of a single room if it makes the house better. He recognized that when we take a system apart and break it down, we cause it to lose all its essential properties. In return, we lose our understanding of how that system works because we can no longer examine the interactions within it. As he said, "Only through design do you deal with the whole and move through the parts." We need context in order to adequately address problems, and context is found within the environment, interactions, and systems.

30

Complexity theory marries
this kind of systems thinking with history.
It's really trying to understand how patterns change.
So we have to ask the questions:
What are the patterns we are currently in
and how might they change?
And how deeply embedded are they?

Jean Boulton
Episode 075 Embracing Complexity

Developing a Systems-Thinking Lens

Our series on design thinking—also known as human-centered design—helped us better understand complexity and systems thinking. We were amazed by the parallels we found between design thinking and complex systems theory. The most interesting similarities we found were in human characteristics, ideologies, and perspectives.

At the core, both complexity and design thinking are utilized by people who keep an open mind and hold space for new perspectives and ideas. They practice expanding their points of view by making an effort to, at a minimum, acknowledge other perspectives. Complexity thinkers and design thinkers are curious people and their non-linear thinking is driven by emotional intelligence. There isn't room for ego, stubbornness, or rigidity when evaluating systems. Using a systems lens requires tailored behavior, a fine-tuning of self, in order to gain clarity.

Self-awareness brings more attention to our mental models, which might be holding us back and keeping us from seeing the big picture in an unbiased way. Many of us fall victim to mental traps. The trick is to recognize when it happens so you can shift your perspective, because those who can observe and engage without attached bias are able to identify connections as they emerge. They see beyond just what they want to see. They don't judge or make assumptions. They don't force their will upon any system. Instead, they listen, not just to what people say, but how they say it. They question everything.

Finding meaningful connections can strengthen any system or design. In a presentation given by Stephen Kaisler and Gregory Madey at the HICSS-42 conference, "Complex Adaptive Systems: Emergence and Self-Organization," the authors shared, "The ways in which the agents in a system connect and relate to one another is critical to the survival of the system."

We continue to be fascinated by the connections we discovered in our journey to learn at *Human Current*. We don't claim to be experts on any one methodology, framework, or theory. We just loved the opportunities that emerged from our interactions with each other when we consciously held space and appreciation for each other. We heightened our consciousness with every casual conversation, interaction, and exchange of ideas. It's a practice which will continue to help us develop our systems lens so we are better able to identify the connections within our human systems—the systems that affect us all.

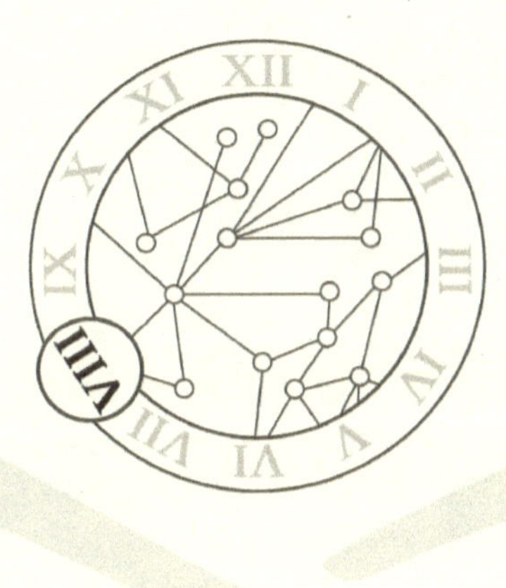

AUGUST

1

Sharing Successes

Doug has been an amazing mentor for so many reasons, one of which is for the nontraditional business practices he taught us. For example, rather than having weekly or monthly all-staff meetings like most companies have been doing for decades, Doug insisted that we practice sharing successes. His sharing success model simply requires a team leader to hold short, informal meetings with team members to ensure they are working at optimal creativity and adding value to the company. Doug taught us that by sharing successes, teams will maximize opportunities for increasing energy levels and creativity.

These are Doug's six steps for practicing sharing successes:

1. Team members share current and recent successes.
2. Team members share their new learnings.
3. Team members explain the next challenges.
4. Team members listen, pay close attention, and hold space for each other, positively and effectively.
5. Team members explain their next steps and action items.
6. The entire team experiences and enjoys the success.

According to Doug, it is the most cooperative who thrive. If this process is done authentically, it generates cooperation, support, appreciation, and help from other team members, building connectedness within the team.

2

Reclaiming Leadership for the Human Spirit

We continued to be in awe—truly humbled and honored—to interview so many extraordinary people for *Human Current*, and Episode 090 Reclaiming Leadership for the Human Spirit was no exception. We interviewed best-selling author, speaker, and teacher Margaret (Meg) J. Wheatley. "My own work now is not in trying to shift the paradigm, but trying to wake up a few devoted, dedicated people to be leaders for this time, which means being warriors for the human spirit," Wheatley said.

She told us she is driven by her "unshakable conviction that leaders must learn how to evoke people's inherent generosity, creativity, and need for community." Wheatley also described how leaders can experiment with complexity theory and systems thinking to better understand the role of emergence and interconnectedness in their work. She said, "Claiming leadership, for me, is a conscious choice to step forward with courage, with a stable mind and in a community of other warriors so we can be a peaceful, thoughtful, discerning presence for others." Wheatley offered a powerful and thought-provoking message for courageous leaders of this time, calling on them to "give up the paradigm of command and control, of treating people like machines, of denying the fact that people are creative and have deep inward motivation."

Wheatley is a management consultant and systems theorist. She applies theories of change and chaos theory to organizational culture and leadership. She described what interconnectedness has to do with conflict, "Because we are all interconnected, we are seeing things differently, so of course we're going to be in conflict." Yet, "When people work within the complexity paradigm, you understand that life organizes without control. There are dynamics and processes that lead to what Stuart Kauffman, the great complexity scientist said, 'you get order for free.'"

3

*We need to continually open up our options to possibilities,
and it's the possibilities for what we don't necessarily expect
and what we cannot necessarily plan for . . .
it is the surprises that allow us
to move in a different direction
and may even give us a jump-start
on developing a new product
or thinking outside of the box.*

Denise Easton
Episode 117 Influencing Pattern-Based Change

4

Can We Design the Right Conditions for Creative Teams?

Creating a culture of creativity requires a systems thinking lens. Like any complex system, creativity is sensitive to environment, it is not a linear process, and it is generative. Inspiration is pieced together from an amalgamation of conscious and unconscious processes, frames, and external stimuli. We stand on (and rub against) the shoulders of everyone in our cultural and intellectual networks.

It's said that creativity comes from the mind, yet thought leaders have taught us to ask questions like, where is the mind? Is there any singular, original idea? If we're finally acknowledging that creativity involves some level of collective intelligence, how should we build teams to yield the kind of creativity that becomes innovative for the organization?

An article on the *World Economic Forum* website, "How Should Leaders Build Creative Teams?," explored creativity as a growing discipline and examined the complex relationship (and differences) between intelligence and creativity. Lee G. Bolman and Terrence E. Deal acknowledged complexity within organizations in their book *Reframing Organizations*, advocating flatter structures and a four-frame model for team building based on informal roles that emerge. John R. Katzenbach and Douglas K. Smith's *The Wisdom of Teams* discussed high-performance organizations, citing shared leadership and self-organization as keystones to innovative teams. The common factors in all cases of the ultimate creative dream team seem to be adaptability, social leadership skills, and a supportive culture.

Ultimately, creativity itself is not "built in" to teams, because you can't build for it. However, you *can* structure for emergence and for tolerance of uncertainty, which is where so many teams get stuck, trying to avoid or achieve specific outcomes. It is important to understand how your team or project works with emergence.

5

Performance is often hard to measure but success, because it is distributed and it's about the community, is much easier to measure, much easier to quantify.

Albert-László Barabási
Episode 115 The Network Science of Success

6

A Philosophical Look at Company Culture

For Episode 038 A Philosophical Look at Company Culture, we shared our interview with Benjamin Taylor, a systems thinker and learner. "Most of my work is really helping people to do transformation or 'serious change' in public service," stated Taylor. He discussed the intended and unintended consequences of leadership and said that "meaning only ever emerges in context." Taylor described the philosophical side of company culture, as well as how leaders can influence company culture. "Most people in this world should be trying to work themselves out of a job." He also shared a quote attributed to Millard Fuller, "It is easier to act yourself into a new way of thinking than it is to think yourself into a new way of acting."

7

Are Complex Systems a Higher Form of Intelligence?

In everyday life we tend to think of intelligence in terms of hierarchies, with the smartest people at the top. We measure IQ scores and make reverse-coherent stories about the qualification of our leaders based on our perception of their intelligence and influence. When systems fail to deliver the results we expected, leaders take the brunt of our blame.

When we consider big-ticket issues like politics, disasters, economics, climate, and inequality from a polarized perspective, aren't we putting too much stock into the agency of an individual? Many paradigm changes seem to be the result of systemic shifts that have a single personality at the crest of a wave—but in reality, there were movements behind those individuals that already had momentum as a network. Where there's drama, it's easy for us to find causality through its players, yet even our myths tell a different story. Take Shakespeare's *Romeo and Juliet*, for example. It was a series of interrelated events and feedback loops—"Misshapen chaos of well-seeming forms!"—and misplaced narratives woven through a network of families that brought about the downfall of the star-crossed lovers.

As we band together to exchange the stories that become our shared realities—retelling, reinforcing, and reacting to them through feedback loops—we have to wonder if those networks of stories are not a better measure of our collective intelligence than the people at the top shouting the loudest. Although we like to say we're at the top of the food chain, in all nature's complexity, we human beings are somewhere in the middle of nature's hierarchy. Structural information precedes our understanding of matter. Our cells do not understand our life plans, and yet they function just the same; likewise, we do not always understand how the larger systems we create and participate in affect us individually.

The key to our evolutionary resilience is perception—our precognitive, sensory ability to recognize complex patterns and discern higher orders at work. What's more, our computing power has now advanced to the point where we can make sense of billions of data points. Networks and ecosystems are beginning to have a voice and tell stories that transcend our individual politicking. Can we hold back on reductionist reasoning long enough to find truth in emergence?

Consciousness, as we know it, is not a requirement for a complex adaptive system to do its thing on some pretty impressive scales. As we have looked at higher powers than our own individual means and, not understanding them, called them angels, perhaps we are beginning to perceive new patterns of collective intelligence at work through the systems we have created, which live nowhere and everywhere in physical space, and still somehow affect all of us.

8

*Some of the responsibilities of leaders
are shifting to teams, because teams are better able
to handle complexity at multiple scales than a single leader.
So, I think maybe the role of a leader is increasingly shifting
to facilitator and in being able to help manage
the flow of information through a team.*

Stacy Hale
Episode 050 Complexity Science Is Everyone's Science

The Complexity and Chaos of Creativity

How does chaos influence creativity? How can "flow states" help teams manage feedback and achieve creativity? In Episode 082 The Complexity and Chaos of Creativity, we talked with designer, educator, and author, Jon Kolko. Kolko shared ideas from his book, *Creative Clarity: A Practical Guide for Bringing Creative Thinking into Your Company*, which he wrote to help leaders and creative thinkers manage the complexity and chaos of the creative process.

Kolko said, "Framing is the way in which the problem is structured and presented and the way that those constraints start to manifest as an opportunity statement." He explained how the elements of complexity science–emergence, constraints, feedback, and framing–influence the creative process. "A constraint emerges from the creative exploration itself. These constraints become a freeing way for creative people to start to explore without having rules mandated at them," he said.

He also shared helpful tips for how to foster a culture of creativity within an organization. We enthusiastically agreed with Kolko's statements: "The rules around trust need to be articulated," and, "Chaos is the backdrop for hidden wonderment and success."

10

If we all became more reflective,
we would become much better equipped
to be of service to this time,
and we would be much more content.
It's the franticness, the rushing around
and the withdrawal from one another
that is really creating this disastrous time
for the human spirit.

Meg Wheatley
Episode 090 Reclaiming Leadership for the Human Spirit

What is Control? Do Leaders Have It?

by Benjamin Taylor for *Human Current*

Control is probably an essentially contested concept. This is to say that not only does it mean many different things to different people, these meanings are grounded in concepts of rationality, human communities, and practices, which may make it impossible for proponents of different meanings to really communicate with each other. But we can try!

We tend to want to talk about one form of control as bad, evil, crushing, or mechanistic. This is the mindset which systems thinkers sometimes liken to being trained to throw stones—reductionist control. If you're good enough at it, and you know the weight of the stone, the strength of the wind, etc., you can predict where the stone will land. But if you're suddenly presented with complexity, say in the form of a live bird, this sort of control can only apply known thinking. So you make the live bird more like a stone, which prevents the nasty habit it has of behaving unpredictably when thrown. It does lose certain essential, birdlike qualities—but it becomes as predictable as a rock (because, now, it's much more like a rock . . . or potentially tied to it).

Then there's another type of control, which we desperately want—a control which is somehow creative or enabling. Another way of putting this is the distinction of two types of power in Scandinavian or Germanic languages (which I learned from Jonathan Horwitz, a very good teacher). So you have *macht*, or might, power over. And *kraft*, or the power to create, power with. Forget about arguing over words, if you can. What do we want control to mean?

I bring it down to two other words—responsibility and constraints. I don't think you can have freedom or creativity without them.

Organizational leaders might not have as much control as they think. For example, sometimes I work with chief executives who are not amongst the most influential in their organizations, but they absolutely do have responsibility. For me, that responsibility comes down to accepting that the results that come out from their behaviors and the way the organization affects people—and peoples' response to those things—are the results they have created. If you really want to lead, first find out what you are responsible for, the actual results. If you can deal with that—confront yourself with that—you've made the first step. If you can learn from it and try to change, you're on the way.

There are also constraints. In any organization, it's about giving people clear boundaries to their work and freedom within those boundaries. There's room for improvisation and instant variation in these constraints, once you're in the swing of it—but freedom without these constraints, freedom which doesn't take account of experience, capability, relationships with others, knowledge of the work to be done—well, that's meaningless. Note that this absolutely doesn't rule out self-organization, collective organization, but it reveals that any of these, to be effective, must have a control system embedded somewhere—in collective agreement, in rules and procedures, or even in the work itself. If leadership is to be meaningful, the leader has to accept responsibility for the emergent outcomes, for the experience of the people in the organization and outside the organization, for the whole thing.

Excerpted from a blog previously published at human-current.com/blog on September 2, 2016.

12

Learning is not just that something is interesting to study, but it is actually a goal. . . . You want to be a lifelong learner and that's true for you but it's also true for your teams and that's an explicit decision that you have to make and something also that you have to explicitly nurture.

César Hidalgo
Episode 085 How Networks Learn

13

The Conscious Evolution of Leadership

We talked with author, public speaker, and co-founder of Wisdom Works, Renee Moorefield, for Episode 086 The Conscious Evolution of Leadership. Moorefield discussed her passion for systems thinking and how she uses this way of seeing the world to help others thrive as living systems and leaders in the world. "Human vitality—our thriving—should not be an afterthought, it really is an asset and it should be a priority in how we are designing systems, our organization, our communities, et cetera." She explored how to nurture our capacity to thrive through self-awareness and holistic thinking practices, since "thriving is an innate capacity within us."

14

Complexity thinking is the thinking side of holistic science. It's that kind of thinking that brings you closer to the way nature really is.

Stephan Harding

Episode 112 Gaia Theory, Holistic Science, and Deep Ecology

15

Using Data and Design to Empower Teams

We talked with innovator and vice president of marketing at iCharts, Rico Andrade, for Episode 049 Using Data and Design to Empower Teams. Andrade discussed the importance of a leader's role in empowering teams and creating the right environment and conditions for them to interact and make real-time decisions.

He said, "Sometimes we overvalue the significance and contribution of a strong, single leader. A lot of leadership is about creating the environment and conditions, in which different parts of an organization can interact with each other and lead themselves without having a traditional hierarchy."

Andrade also explained that because the world is more complex and data-driven, it is becoming more important than ever for leaders to release some control and allow for emergent insights to play a role in guiding business decisions. Andrade argued that "using data as a team member" is a powerful way for leaders to design employee collaboration. He went so far as to explore the ways leaders influence the vision of an organization. "In a sense, what the role of a leader becomes in an organization is more than just setting a vision, but creating the conditions for the vision to set itself in some situations."

16

Making Change That Matters

by Benjamin Taylor for *Human Current*

"If you want truly to understand something, try to change it."
Kurt Lewin

I started off studying philosophy. However, I also studied politics and was involved in politics from a very early age, leafleting with my mum from the time I could carry a bundle of leaflets and joining a political party at the ripe-old-age of sixteen. Back then, I believed that policy was the main driver of change, and even though I edged more toward social movements and ideas at university, I'm glad that I had two experiences which started to knock that idea out of me. Straight out of uni, I had two very operational jobs—admin for a youth development charity and coordinator for an advice center. So I saw something very different—the critical importance of operations to make a difference in the world—the primacy of the quality of service and the experience of the organization, both for the customer (or citizen) and for the employee.

Still, when I got a fancy job as adviser to Leader of the Council (or adviser to the Mayor—the title changed a bit, but the bloke I worked for stayed the same!), I was still convinced that politics made a difference. And it did—but I also saw the compromises, and more importantly the many, many times when *not* acting was the only way to reduce risk and maintain power and influence for the future. I was a non-political council officer, but working directly for political leaders with a thin majority, both in the electorate and within the ruling group. It was paralyzing.

Still, when I moved on to a fancy consultancy (PricewaterhouseCoopers, as it was then called), I still thought that the power of my intellect and insight could change the world. Working with a sublimely talented, competitive, and capable team of colleagues in an organizational structure I ultimately wasn't comfortable with, we produced brilliant reports. Clients were (sometimes) duly impressed. However, it slowly began to dawn on me that nothing was actually changing. Either we were hired to perform some organizational 'pantomime'—a great display of effort for change, without the actual pain of change—or the intentions of our sponsors just weren't enough to mobilize change. We did some great work, don't get me wrong. But more often, we produced high-quality shelfware.

The rest of my career has largely been a downhill spiral. The reports got shorter and shorter, the action plans rose to prominence, rapid improvement events were the focus, then the action plans faded a bit too and we started to primarily just focus on real action—have a look, make a hypothesis (predict something that will improve things!), make a change, regroup and learn, and do it all over again. And keep doing it. Some dismiss this as tinkering—not theory-led or vision-led or big-bang enough. But I've never learned more about the actual organizational systems, and never gottem bigger change done, than now. We still sometimes write reports—sometimes nice ones—and I get to talk about big ideas. But the real work is making change that matters. So when I was asked what action, what next step, I would recommend, I jotted down: Learn! Explore! Think! Have a laugh.

The final step is doubly important—a huge risk in all of this is coming to take yourself too seriously, and coming to lose your joy in the work. And you couldn't learn more, and make a bigger change for the better, than making an experiment on your culture: in your organization, try to find the stories that encapsulate the gut reactions people have had to organizational systems, symbolism, and leadership behaviors. Try to forecast what change in the latter would change the former positively. And predict what the response will be—and how you'll respond. If it's in your power, do it. And see what happens.

Excerpted from a blog previously published at human-current.com/blog on September 2, 2016.

17

Because businesses are so complex today,
leaders really need to understand
the notions of complexity, emergence, tipping points.
And even though tools like statistics are still useful,
they need to understand what the limitations are.

Paolo Gaudiano
Episode 048 The Intersection of Diversity, Innovation, and Performance

18

Visionary Leadership in a Complex World

"Solutions often emerge on their own with a little coaxing," said consultant, blogger, and systems thinker Scott Jancy in Episode 047 Visionary Leadership in a Complex World. Jancy reflected on his background in architecture and discussed why he believes design and leadership are synonymous. He explained how leaders can successfully utilize systems thinking for problem-solving: "Innovation comes when making small connections and linkages between things as you're spending hours doing the same thing over and over in support of a vision." Jancy emphasized the importance of leadership vision while approaching complex problems: "Complexity is easy to deal with when you know what you want and can figure out what's best for you."

19

Complexity Science Is Everyone's Science

For Episode 050 Complexity Science Is Everyone's Science, we interviewed professor, complexity scientist, and founding president of the New England Complex Systems Institute (NECSI), Yaneer Bar-Yam. Bar-Yam talked about how we can understand complex systems science by applying it across all of the systems we interact with in our society today. He also explained the importance of flatter, team-oriented organizations when dealing with complexity because today's hierarchical organizations are limited by "what one individual can do and individuals have a limited degree of complexity that they can cope with."

Bar-Yam said, "When you study complex systems more carefully, you realize that having a strongly connected system will actually limit the ability of the system to have complex behaviors. So it turns out that really complex systems actually have a structure that has different levels of organization, so there are small groups and then larger groups that include the small groups, and yet larger groups, all the way up to the system as a whole."

Our other guest for this episode was our former co-host, Stacy Hale. Hale shared, "What Yaneer taught me, that blew my mind and pivoted my thinking about complex systems, was that there is a trade-off between scale and complexity." Hale pondered, "How do we understand complex systems science in a way that we can apply it across all of the complex systems that we currently touch in our society today?" She observed, "The role of leadership changing as society becomes more complex."

20

Become a History Buff

Over the years, we've learned that people's level of systems thinking varies across a wide spectrum. We believe that becoming a history buff is a helpful approach to becoming a better systems thinker.

Learning about history can be interesting, surprising, and at times, even disturbing. Often there are different theories and perspectives on how things have come to be. For us, learning about various theories or perspectives—for example, the history of societies, religions, the brain, the environment—has opened our minds and inspired us to learn more and more.

History makes up the narrative of how people, places, and things have gotten to where they are today. Knowing the history helps us understand patterns and trends, which is an important practice for being a systems thinker and is a skill you can transfer to many different areas of your life. We have applied our knowledge of organizational communication and learning about the history of people, places, and things to our work with *Human Current*.

21

So, in simple terms, your performance is about you. Your success, however, is about us.

Albert-László Barabási
Episode 115 The Network Science of Success

22

An Exploration of Social Minds

At the International Conference on Complex Systems 2018, Simon DeDeo said, "The hard problem about creativity is, of course, the origins of it." We shared our conversation with him in Episode 108 An Exploration of Social Minds. DeDeo is assistant professor in social and decision sciences at Carnegie Mellon University and external professor at the Santa Fe Institute. He talked about his research, which explored extremely creative people and events throughout history in order to learn where new ideas come from and what people do when they get them, describing his research as "an alternate account of what it means to be human." DeDeo explained that we are not infinitely predictable creatures; rather, we harness an extreme capacity to evolve and create. "It's often said that evolution is the most powerful force for destruction, but also creation, that we have ever seen. . . . One of our goals is a sense of wonder."

23

The Complexity and Chaos of Creativity

Are chaos and creativity synonyms? Is creativity an emergent property of chaos and complexity? The concept of creativity resurfaced often on *Human Current* and its usefulness in dealing with complex adaptive systems. We learned that creativity requires trust and freedom. It also requires stepping outside of established patterns while embracing uncertainty, vulnerability, and chaos. It can help us make unexpected connections and push boundaries into the unknown. We can work toward expanding our creative abilities by leaning in to our imaginations. We can let go a little, experiment often, quiet our minds, and seek childlike wonder and curiosity. Here are some of our favorite quotes for capturing the complexity and chaos of the creative process:

> "Creativity requires the courage to let go of certainties."
> Erich Fromm

> "Creativity is just connecting things. When you ask creative people how they did something, they feel a little guilty because they didn't really do it, they just saw something. It seemed obvious to them after a while. That's because they were able to connect experiences they've had and synthesize new things."
> Steve Jobs

> "Creativity involves breaking out of established patterns in order to look at things in a different way."
> Edward de Bono

> "You can't use up creativity. The more you use, the more you have."
> Maya Angelou

> "Creativity is the ability to introduce order into the randomness of nature."
> Eric Hoffer

24

The Network Science of Success

"The key for success is to find a happy moment in whatever you do and don't rely on external recognition to justify your choices." Albert-László Barabási, professor at Northeastern University where he directs the Center for Complex Network Research, said this in our conversation for Episode 115 The Network Science of Success. Barabási, whose most recent book is *The Formula: The Universal Laws of Success*, revealed that he changed many aspects of his life "thanks to the research that went into this book." He shared key takeaways and important lessons from his research on the science of success. "One of the things I learned writing *The Formula* and doing the research is how important persistence is. If I look back at my life, one of my biggest assets was that I believed in my own ideas and often I was willing to pursue them for four, five years at a time without any sign of success."

He described his work this way: "What I do on a daily basis is figure out how networks affect our lives and lately I am very interested in how networks affect us as individuals and how they determine our success." Barabási discussed how he defines success, including his own. He said, "Success and your ability to succeed in a given environment is partly affected by networks, but also affected by other things like your performance, like your ability to adjust to your environment. So in the end, the research and consequently the book itself became not only about networks, but about success in general."

Barabási said that success captures a community's reflection on a person's performance, and shared that he asks himself: "What does the community see and how does it appreciate my performance?" He added, "If I can convey the mechanisms through which success and reward are generated by the community and how you can turn your performance into success, into something the community can benefit from, then I achieved my goal with *The Formula*."

25

Seeing the Big Picture

Being able to zoom out to see the big picture or whole system and, at the same time, understand the smaller parts that make up the system is a characteristic of a systems thinker. We appreciated how Martin Linsky and Ronald Heifetz's book, *Leadership on the Line*, described "learning from the dance floor and the balcony" to see the big picture.

Learning the balance of when to be on the dance floor or the balcony can take some practice—and it's well worth it. This discipline helped us at work, with friends and family, and even when parenting. As Linsky, Heifetz, and their co-author Alexander Grashow said in *The Practice of Adaptive Leadership*, "Knowing how the environment is pulling your strings and playing you is critical to making responsive rather than reactive moves."

When working with organizations, seeing the big picture has traditionally involved understanding the various structures or departments that make up organizations, the people within those departments, and how they collectively make up the whole. We would argue that seeing the big picture also involves understanding the complex relationships between the people in the system, the departments, the organization, and their extended community and society.

26

AUG

The data science era tends to look
at people as samples from a bag.
Our models of traffic, for example,
these are things that produce really compelling accounts
of why we get traffic jams and
of course what they do is they have these
very simple agents who sit in cars
and they accelerate and decelerate in certain patterns.
So, we've had a lot of success with that approach.
For us, there's a limit to that because creativity
is often a tail phenomenon.
It's an extreme phenomenon in many ways
and so one of the things that we do is
we go look at those extremes.
We go to points in time, we go to civilizations,
we go to people who we believe might be
sources of great human creativity.

Simon DeDeo
Episode 108 An Exploration of Social Minds

27

Shifting Mental Models

We all have them, whether we know of them or not. "Mental models are deeply held internal images of how the world works, images that limit us to familiar ways of thinking and acting. Very often, we are not consciously aware of our mental models or the effects they have on our behavior," wrote Peter Senge in *The Fifth Discipline*.

Mental models can get in the way of our success and even of ourselves—our best selves. The good news is, once we are aware of our mental models, we can begin to shift them. In order to shift our mental models we often have to do some unlearning. The same is true for company culture; sometimes a company has to unlearn ideas, beliefs, attitudes, and even behaviors. Whether you are interested in shifting your mental models or your organization's, there are a variety of resources available to help, including Senge's famous Ladder of Inference, which makes explicit the inferential steps we take from available data to a decision and action. Shifting our mental models has helped shape us as systems thinkers. As a result, we have deepened our ability to see situations from different perspectives and increased our understanding of the situations in which we find ourselves.

28

Preparing Leaders for Complex Change

For Episode 121 Preparing Leaders for Complex Change, we talked with chief executive officer of Toffler Associates, Deborah Westphal. Westphal shared how her background in engineering helped her as a systems thinker. "Being an engineer by education, it really is a necessity to think as a system and to understand the system of systems and understand those interrelationships."

Westphal explained four macro-drivers that cause uncommon disruption and influence everything we know about organizations. She explored very important questions and assumptions about power structures, technology, and societal values. "Some of the problems we are faced with—climate change, clean water, and food—are going to take deep coalitions to solve because no one entity owns the problem. But there are so many entities that are a part of the solution. We've got to come together in uncommon ways to solve some of these problems."

Westphal advocates for leaders to focus on people, rather than processes or technology. She said, "The Information Age is not about the technology. The technology has fueled it, but it's really about the changing nature of people." Westphal continued to say, "Once technology is developed, people are going to change, and relationships are going to change, and policies are going to change, and regulations are going to be debated. So understand that everything really is connected and nothing changes in isolation. You have to isolate the problem, but then to solve it, you've got to go big and then you've got to go small again to get the real answer." She pointed out, "Nothing happens in isolation."

29

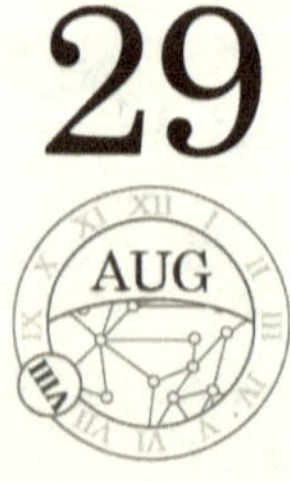

Wonderful though science is, it is very one-sided.
It focuses entirely on what we can discover about nature
using our rationality, our reason, which is of course
very important and we don't want to give that up.
But, we also recognize there are other ways
of knowing about nature
which don't so much focus on reason.

Stephan Harding
Episode 112 Gaia Theory, Holistic Science, and Deep Ecology

30

How Mindfulness Impacts the C-Suite

Most C-suite executives spend their days on their toes, managing a great deal of complexity. The morning may start with a board meeting, then a check-in with upper management, a review of reports, then a lunch meeting with a concerned client, and so on. With such an intense daily routine, executives are always in action. Unlike athletes, they rarely get the chance to recharge physically or mentally. When CEOs are burned out and no longer find themselves at the height of mental clarity they need to make sound decisions on a daily basis, they often turn to outside help, like Pilar Angel. As a co-founder of Axeos Performance Institute, Angel champions meditation and mindfulness exercises. She supports her clients in embracing a new set of activities that add rocket fuel to their work.

When we interviewed Angel for Episode 026 Mindfulness is Complexity's Best Friend, she shared the RAIN tool with us. The acronym RAIN was first coined about twenty years ago by Michele McDonald and is further described in an article by Tara Brach, "Feeling Overwhelmed? Remember 'RAIN'" for *Mindful* magazine. It's an easy-to-remember tool for practicing mindfulness, with four steps:

Recognize what is going on.

Allow the experience to be there, just as it is.

Investigate with kindness.

Natural awareness, which comes from not identifying with, or judging, the experience.

31

I think that if people go through the world
understanding that certainty is an emotion,
then we can trust it less.
It's a great way to feel,
but it actually inhibits us quite a lot.

Jennifer Garvey Berger
Episode 125 Learning How to Thrive in Complexity

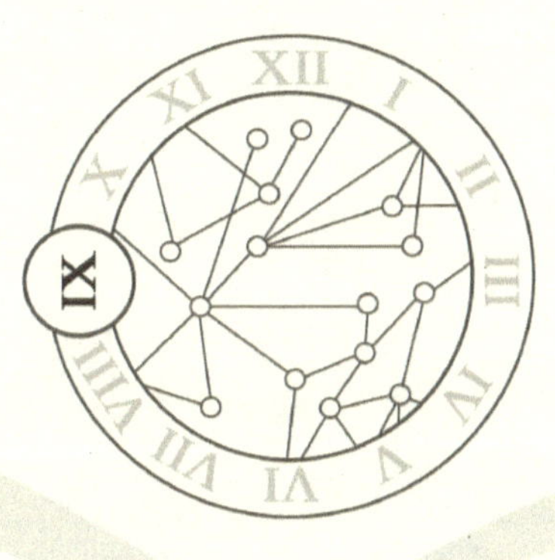

SEPTEMBER

1

Doug's Gold Standard

We had been learning from our mentor, Doug, for years, so it was only fitting that during our series on education we asked him about his most radical learning or educational experience. We shared Doug's response in Episode 010 Super Education, "I attended several Ivy League schools, but my most radical learning experience was while working at the drugstore when I was twelve years old, making sure people were taken care of in a beautiful way, which is really all you do in business anyway."

Many years after Doug worked at the drugstore, he developed his own gold standard—a model of excellence. It was very important to Doug that we all strive to uphold the gold standard of listening, note taking, learning, and total immersion. Here's a quick look at Doug's gold standard:

The gold standard of **listening** is to be able to hear something one time and get it. Get all the words, concepts, and "listen with full being," Doug would say.

The gold standard of **note-taking** is to be able to document your activities and meetings, and commit them to memory. Absorb everything that is said by listening and writing the exact words and meanings.

The gold standard of **learning** is to see once, do once, and teach once. This method locks in the learning.

The gold standard of **total immersion** through peer review and peer learning brings the learning to life. Practice a peer review by quickly pinpointing related online information and literature, and simply jump into what you discovered. Next, discuss learning with peers to get other opinions and feedback for total understanding.

Another method for total immersion learning is mentoring. Find a mentor for yourself, and mentor someone junior to yourself.

2

A living system is a learning system.

Meg Wheatley
Episode 090 Reclaiming Leadership for the Human Spirit

3

How to Be Your Own Résumé

Some people believe in a linear career path. Others stop waiting for permission and create their own paths. Entrepreneurial thinkers understand that they are part of a network, not a link in a chain. So why should we treat higher education like a conveyor belt? Decreasing transaction costs in education, plus rapid technological advances, are making it possible to experiment with being a renaissance person *and* land an incredible job.

In Episode 011 How to Be Your Own Résumé, we talked with Isaac Morehouse, founder of Praxis and a writer who is obsessed with human freedom, education, and entrepreneurship. Praxis is a one-year program where you learn by doing. Participants work alongside founders and chief executive officers while also completing a rigorous educational experience which includes one-on-one coaching, self-guided projects, hard and soft skills training, and more. Why? Personal mastery is more than being a subject matter expert in a specialized area. It's knowing how to weave together a vast network of skills, passions, people, and ideas.

"The idea that you can achieve expert status in a classroom and then step into the market is silly," Morehouse said. "I think the dichotomy of working and learning will continue to blur." Praxis does just that.

At Praxis, focus is self-directed by the participants and includes philosophy, logic, ethics, economics, digital skills, management, history, culture, and more. The program is producing a workforce of renaissance people and polymaths. At the end, graduates present a portfolio project on a website they've built and demonstrate their mastery through podcast-style conversational interviews.

"Creative problem-solving requires an understanding of the complexity around us, and knowing our place within that system, and where we can create value," Morehouse said. "Our market needs complex thinking based on simple rules, not simple thinking based on complex rules."

4

Decentralize and Unify

Boston College professor Patrick McQuillan talked with us about his work in education reform and urban school leadership for Episode 098 School Reform with Complexity Theory. McQuillan shared why it is so important for school leaders to decentralize control and work toward a unified school culture. He said, "One of the things that probably every third grader knows is power should be decentralized. People need to have power in schools. They need to be able to act. They need to be able to make the school an adaptive system." He described "the thread of adaptive leadership" as a productive approach to systemic change.

5

Complexity Live

We were honored to collaborate with Complexity Labs for a series called *Complexity Live*, which were streaming video conversations about all things complexity. Anyone with a curious mind could listen in and ask questions while the conversations unfolded. The *Complexity Live* group discussions included up to seven people and we extended invitations to anyone interested in being an active participant. The conversations covered topics such as systems change, building networked organizations, systemic risk, and VUCA (volatility, uncertainty, complexity, ambiguity). As the moderators of the *Complexity Live* discussions, we held space to explore new ways of thinking. The recordings are available on the Systems Academy *YouTube* channel (formerly Complexity Labs).

6

I feel that it is our responsibility as scientists to translate what we do to a wider community.

Albert-László Barabási
Episode 115 The Network Science of Success

7

The Power of Substructures

We felt fortunate to be introduced to Yaneer Bar-Yam by our former co-host Stacy Hale, who attended the winter school program offered by the New England Complex Systems Institute (NECSI). Hale described NECSI as "the leader in producing the research that tells us how complex systems science can be applied to real-world challenges, specifically challenges dealing with policy decisions and how they have unintended consequences that can happen globally." She shared her learning experience, which we aired in Episode 050 Complexity Science Is Everyone's Science along with our conversation with Bar-Yam.

Bar-Yam has the ability to explain complex systems science so gracefully that even the most inexperience complexity thinkers can easily tackle the principles, methods, and concepts he presents. Bar-Yam clarified some common misconceptions about complex systems for us. For example, he said, "When you study complex systems more carefully, you realize that having a strongly connected system will actually limit the ability of the system to have complex behaviors." We were surprised by this statement, because we were admittedly guilty of assuming that more connected networks and systems would be more powerful than those which have fewer connections. It turns out the structures of complex systems are . . . well, complex. We learned that a system's power often lies in the way in which it is organized, including the way its levels and substructures interact, and the function of the system itself will often dictate its structure.

Your brain, for example, is a complex system with millions of neurons that connect with each other, but there are also many subsections in the brain. The human brain appears to be split into two sides and functions within it are segregated into substructures. This design makes our brains more powerful than they would be if they were fully connected networks where every neuron is connected to every other neuron. This highly functional design illustrates why Bar-Yam advocates for teams and networks, calling them the ideal structure for managing or working within complexity.

8

What we haven't yet done
with the basic science of complexity
is really make it extremely clear
that it's an underpinning of what people should learn.

Stephen Wolfram
Episode 101 A Journey of Computational Complexity

9

Super Education

For Episode 010 Super Education, we had a casual conversation about education and explored radical educational experiences. Co-host Angie Cross stated, "School helps us realize what we don't know." Co-host Stacy Hale described how values and mental models can influence our learning. "We all walk around running different value systems and different mental models and we filter our learning through those perspectives." We explored different perspectives on learning and discussed how "real learning gets to the heart of what it means to be human," a quote from Peter Senge's book, *The Fifth Discipline*.

We also discussed how Doug would describe his most radical educational experiences working at a pharmacy when he was just twelve years old. The following is an excerpt of us discussing his learning:

Cross: Jim, Doug's boss [at the pharmacy] was less about the hierarchy and authority and more about connecting with Doug around: What is your experience? What are your values? And what did you learn today? I can only imagine this young twelve-year-old Doug excelling in that environment, and that being some foundational learning, influencing the Doug that we know and love so much today, that has taught us so many things.

Hale: He has gone on to influence us in how we learn and create. So his work at the pharmacy worked its way into his value systems and then into ours. So people can make a difference just by learning in certain ways, that's pretty crazy to think about.

Cross: Does that make us all teachers?

Hale: I think we're always teachers and we're always learners at the same time. It's cool to think about. We're going to be looking at the changing landscape of education and all of these different models, alternative models of education and how they are affecting the workforce of the future and, in the true terms of complexity, how the changing workforce is affecting education, because they co-evolve to suit who we're becoming as a people. And the traditional models of education are changing.

10

The superpower of all superpowers is learning.

Rick Hanson

Episode 064 Our Evolving and Complex Human Brain

11

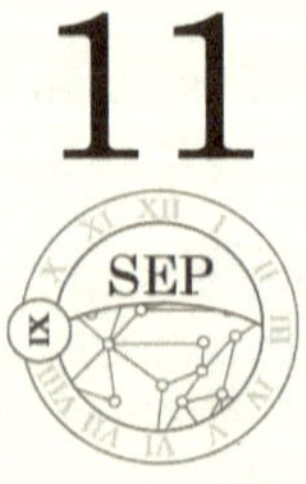

Metanoia, Emergence, and Systems Change

When the hero's journey begins, there is metanoia. The spark that begins the quest, the change in mind or heart that drives the search of the mystic. In Gnosticism, it's an aeon, which is a name for one of the various emanations of God.

Over the years, we have been wondering about the congruence of metanoia to emergence in complex systems. Emergence is the "eureka" that happens when a system interacts with its own parts and produces something new and unexpected.

Through Peter Senge's work, we deepened our understanding of what it means to be a learning organization. Senge used the term "metanoia" for his work in transforming organizations in a positive way from the inside out and described it as "letting go of control and embracing learning as living." Can we liken metanoia to emergence in a system as complex as the way we learn and grow?

In the gnostic myth, Sophia, the embodiment of divine wisdom, creates something solely from herself and gives birth to a being of ignorance. What happens when divine wisdom and ignorance are face-to-face? When we don't understand something (or ourselves), we create. That desire to understand moves us to learn and grow and to do that, we have to access a part of ourselves that can't be explained away by the books we've read or the laurels collected in our curriculum vitae.

Metanoia embodies the realization that the world has become fragmented from unity into polarity, and that even our selves are fragmented into a collection of masks in which our purest, divine being still resides. In Christianity, metanoia is commonly called redemption. Gnostic priest David Hogan said to former co-host Stacy Hale, "Metanoia is the cause for realization. It's the place in us that we call consciousness and awakening." Hogan explained, "Metanoia is very much akin to redemption, or liberation in a sense. The goal is for wisdom to be so deeply embedded in us that it becomes unconscious, that it moves and flows through us.

When you start to free up your concept of self and your ideas, all of your perceptions and the gravity that they hold start to change."

At the moment we find ourselves impressed by our own knowledge and wisdom, metanoia uncoils and deeper layers of learning are exposed so that still more lessons can be learned. More of our hang-ups are gently revealed and as we shed them, the more diverse our idea of self is free to be. Metanoia can be the moment you are compelled to do something to change yourself so that the world can change with you.

12

Not only do we live in
a knowledge economy, but also an economy
which rewards creativity and innovation,
and often these people either have
alternative backgrounds or come
from alternative schools.

Khotso Khabele
Episode 010 Super Education

13

Systems Citizens

In *The Fifth Discipline*, Peter Senge referred to "systems citizenship" as the leadership mandate for this millennium. We were curious about how we can all be better systems citizens, so in Episode 009 Systems Citizens we asked students of the master of arts in leadership program at Royal Roads University in Canada how studying complexity helped them understand the world and solve real-world problems. Kacem Habiballah said, "The way I look at systems, it is a continuum and the boundaries in systems thinking are set by the discussions that are taking place."

Cathy Sturgeon responded, "Systems thinking requires me to be aware about my own assumptions, preconceptions, the way I see the world, the way I perceive the world, my expectations, and being really intentional about challenging how I look at situations, how I look at systems, how I perceive it. It forces me to be open and flexible, and look at different ways to get to a destination."

14

Personal Growth, Radical Learning, and Alternative Education

Are some learning experiences more influential to our personal growth than others? Why are more people seeking alternative learning programs? These questions got us thinking about our educational background and the journey that ultimately led some of us to enroll in alternative learning programs. Co-host Haley Campbell-Gross shared her educational story:

I have had a diverse educational background. It was mostly standardized, but also very unconventional.

Right before entering fourth grade, my mom pulled me out of public school to homeschool me. She was a very dedicated stay-at-home mom, teacher, and coach. The curriculum focused on writing, grammar, and critical thinking, which were required for most activities, and rather than testing, I was evaluated on comprehension and problem-solving. It was a very intellectually liberating learning experience.

Then, when I was fourteen, my parents decided I should attend public high school. The transition was difficult. Not only was the learning environment totally different, but so was the curriculum, pace, and teaching style. My classes kept moving forward, whether I understood the topic or not. And testing, as well as preparing for tests, become a regular event. I struggled with the concept that my opinions, thoughts, and ideas didn't matter.

Sadly, my experience in community college wasn't too different from high school. I took much of the same courses; they were just a little more liberal. I transferred to the University of Houston and enjoyed business school. However, when it was all said and done, I didn't feel very knowledgeable in marketing. But that's what my degree said, so that's what I paid for.

Two years after college graduation, countless hours of googling how to do things, and some on-the-job experience in marketing, I decided to take a course with General Assembly. They're an alternative education startup and their mission is to "transform thinkers into creators," which really resonated with me. The class I attended gave me the opportunity to grasp the current digital marketing landscape while acquiring actionable skills for my career. It was a great learning experience, offering a good balance of structure and flexibility.

Soon after realizing how diverse my educational background had been, I started to wonder which learning experiences from my past were most influential in my life. And looking back, I noticed that all the formal educational experiences were actually less impactful to my personal growth than a few stand-out, informal experiences. I learned the most when I was thrown into a situation and had to figure out what to do next. Maybe it's just my personality, or maybe it's more common than I think, but when structure and rules are enforced in the learning process, it can limit personal growth. When I set out to learn something in a structured environment, I find myself trying to control or at least predict the outcome. It limits my self-awareness, my adaptability, and my creative thinking abilities because I'm approaching it as "training" and letting the system do all the work for me.

Although I do believe there is a time and a place for this type of learning or training, more radical and impactful experiences are not this structured. They happen when you take responsibility for your own growth and create your own goals, even though you may not know exactly what to do or what's going to happen in the end.

We believe that the future of learning will lean toward alternative education. These offerings will become more and more popular as people strive to acquire skills that they didn't learn from traditional education. There is definitely a demand pulling more unconventional learning opportunities into the marketplace, and the more available they are, the cheaper they will become. So many people are ready for a shift in education. These people question the status quo and aren't afraid to act. They seek more intellectual freedom, less top-down hierarchy, and more authentic connection. This will spearhead a new movement in education from grade school to the professional world.

15

I found a lot of the math that I learned in school really didn't apply to complex networks, you need a different kind of mathematics or a different set of models.

Joss Colchester
Episode 089 Nonlinear Systems and Technology with Complexity Labs

16

Holistic Science

For Episode 087 Holistic Science, we talked with Jean Boulton, co-author of *Embracing Complexity* and a management consultant specializing in complexity theory. She shared details about the only master of science in holistic science program in the world. It's at Schumacher College, where she is a tutor and lecturer. She described holistic science as being "slightly broader than complexity science, but it certainly overlaps," and she explained, "It's broader in the sense of it's really emphasizing the interconnectivity between things. This idea of holism is really embodying this sense of everything being connected." Boulton went on to reveal the unique offerings of this program and the importance of understanding the emergent properties of whole systems through the lenses of chaos, complexity, and Gaia theories.

17

A Conversation with Complexity Labs

We discovered Complexity Labs in 2017 after watching a number of videos on their *YouTube* channel and reading their complexity-related articles. In 2018 we formed a supportive working relationship with their founder, Joss Colchester. We believed Colchester and the Complexity Labs team were making an important contribution to the systems and complexity thinking communities with their easily accessible learning content. Since Complexity Labs was established, they have made a great impact with more than 300 video lessons, which have been viewed more than two million times. On the Complexity Labs website they said, "We see complexity theory as a next-generation modeling framework for the sciences and a key tool for solving complex challenges in the twenty-first century."

In Episode 089 Nonlinear Systems and Technology with Complexity Labs, we spoke with Colchester to discuss his motivation for creating the most extensive online resource on complex systems. He described Complexity Labs as an internet-based social business dedicated to solving problems in the most effective way possible. He explained that most people who come into contact with complexity and systems thinking are academics or work in fields like quantum physics, so he created Complexity Labs to work toward increasing access to these important fields of study.

Colchester told us about the courses offered on Complexity Labs. We were particularly interested in one of their courses, "Blockchain Introduction," which provides an overview of this technology and how it can be applied to society and the economy. Colchester provided us with several different examples of how people and companies might utilize blockchain and why he thinks it is important for the average person to be knowledgeable about it. He explained that this emerging technology is evolving rapidly and holds the capacity to decentralize many industries that we interact with on a daily basis. He thinks "the blockchain is going to be the coordination mechanism of how we, in many ways, structure our societies and our institutions, and that's going to happen quite fast." This was a profound statement, which made us wonder, what does it truly mean to be a vigilant complexity thinker in our rapidly changing society?

Our reflections reminded us that we need to be aware and well-informed about our environment and how it is influencing us. This means not only understanding more about the complex adaptive systems that are all around us, but also understanding the complex technologies that affect our everyday lives, including how we communicate and relate to one another. As the world becomes more complex, global, and integrated, we are being called to practice discernment in every aspect of our lives. We need to arm ourselves with both attuned presence and knowledge to navigate this landscape, which is evolving faster than ever before.

We believe Complexity Labs is a precious resource that has provided a way for us to access knowledge that is digestible and usable in our every-day lives. We give thanks to Colchester and his team for their contribution to this space!

Complexity Labs is now Systems Innovation, and Complexity Labs content can be found there at systemsinnovation.io.

18

I think people aren't afraid of change,
they're afraid of the loss that exists within it,
and the grieving of those aspects that we lose.

Jason Dykstra

Episode 003 Using Systems Thinking to Resolve Conflict

19

Complex Solutions for Complex Problems

For Episode 066 Conference on Complex Systems 2017, we talked with Carlos Gershenson, research professor, editor-in-chief of *Complexity Digest*, and co-chair of the Conference on Complex Systems. Gershenson discussed how the many problems that we face today will not be solved by themselves or with "simple solutions." He said, "We need complex solutions for these complex problems."

He went on to say, "Science has an opportunity to solve these problems in collaboration with governments and society," and that only with full participation across disciplines would there be solutions. Gershenson also discussed the 2017 Conference on Complex Systems, how it is relevant to events happening in the current world, and previewed details about the next year's conference.

20

A Journey of Computational Complexity

For Episode 101 A Journey of Computational Complexity, we interviewed Stephen Wolfram, creator of the computational platforms Mathematica, Wolfram|Alpha and the Wolfram Language. Wolfram talked with us about his professional journey and reflected on almost four decades of history, from his first introduction to the field of complexity science to the thirty-year anniversary of Mathematica. He said, "We just celebrated the thirtieth anniversary of Mathematica, and it's been really nice to see over the past thirty years all the incredible things that have been invented and discovered in the world with our technology in lots of different areas." Wolfram described his story as "basically an alternation of working on basic science and working on technology."

Wolfram shared his hopes for the evolution of complexity science as a foundational field of study. "What we need is a methodology of complexity. Once you have a uniform set of methodological ideas, you have more development of a community." Wolfram also gave advice for complexity researchers, recommending that they focus on asking simple, foundational questions. He said, "It sounds paradoxical, given the name of the field—complexity. But work on simple questions."

21

How You Show Up Matters

The messages that were shared at the WorkHuman 2018 conference were not just for human resource leaders. If we want the workplace to change, we all have to change. It is often uncomfortable and not easy, but rest assured—that uncomfortable feeling is your ticket to growth and freedom. Lean in to this feeling and watch the magic happen. There are small moments every day for us to practice showing up.

- We can listen when we want to talk.
- Share when we want to stay quiet.
- Ask someone a question when we feel like judging them.
- Assume good intent when we want to assume the worst.
- Pause when we want to react.
- Ask for feedback when we are afraid.
- Ask for help when we need it.
- Admit when we don't know something.
- Take responsibility when we want to blame others.

Let's practice mindful living and show up to our lives, even when we don't want to. As Sheryl Sandberg wrote in *Lean In*, "We cannot change what we are not aware of, and once we are aware, we cannot help but change."

Why Innovative Companies Hire Entrepreneurs

Michelle Battle-Fisher said, "It's OK to think outside of the textbook," in Episode 012 System Thinkers Share Their Stories. Today, companies are seeking employees who want to demonstrate personal mastery and are driven by owning their work, and these qualities are typically associated with an entrepreneurial background. It's no wonder "one in three employers said they are looking for entrepreneurial experience in their potential hires," according to Millennial Branding in the "Millennial Branding Student Employment Gap Study."

With the ever growing complexities of today's business world, companies are beginning to search for talented individuals who can think both analytically and holistically. Entrepreneurship is a favored skill-set because it exposes people to a real-world and multifaceted experience, and so does alternative education. In fact, many alternative programs are designed to inspire young people by giving them the freedom to think like an entrepreneur and master skills like resilience, innovation, creativity, and adaptability. These programs encourage students to showcase their skills and capabilities in more active, practical ways that demonstrate their passion and mastery—conversations, projects, websites, and other original works. When you're an entrepreneur, the project is you.

23

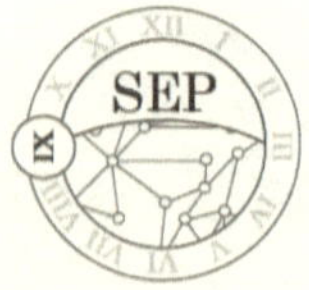

Non-Traditional Learning Programs

The educational landscape is changing and the list of alternative education programs continues to grow every day. Here are a few of our favorite examples of non-traditional learning programs:

- Praxis, a rigorous education and professional development program (discoverpraxis.com).
- Khan Academy, a catalog of online courses offering free world-class education for anyone, anywhere (khanacademy.org).
- The Waters Center for Systems Thinking promotes the effective application of systems thinking concepts, habits, and tools in classroom instruction and school improvement (waterscenterst.org).
- Systems Innovation (formerly Complexity Labs), an open-source, community-powered platform for complex systems research and development (systemsinnovation.io).

24

The Social Pursuit of Order

For Episode 109 The Social Pursuit of Order, we had a conversation with Loren Demerath, author and professor of sociology at Centenary College of Louisiana. Demerath was hopeful that his research would lead to "a model that shows how efficient language development makes a system of agents, and how agents organize themselves by developing a system of language." He went on to say, "We are really looking to create a model of how any kind of agent of information creates a community that eventually becomes bounded based on language. We are trying to evolve language, basically."

He discussed the fundamental nature of complex systems and how he applies concepts of complexity science to his research on culture, social interaction, and the emergence of order. He also highlighted the opportunity to work in a cross-functional way: "Network analysis is all over the place in my field [sociology], and so that's the way we will be connecting with a lot of the physicists who are doing sociophysics."

25

When I think in terms of how intention can guide you into being in the right place at the right time, that has happened so often in my career that somebody could say, "Well, that's like magic."

Dean Radin

Episode 081 The Science, History, and Implications of Real Magic

26

Stories from Network Science PhDs

In Episode 107 Stories from Network Science PhDs, we had a discussion with network science researchers Brennan Klein and Sarah Shugars. Both were network science doctoral students at Northeastern University, yet they shared very different stories for why they decided to study network science and how they plan to approach and apply their research.

Klein said his studies were broadly "somewhere around the intersection between networks and complex systems," and for him, that is looking at "a lot of complex processes that happen all around us." Klein studied cognitive science as an undergraduate and conducted a simple experiment where people at the bottom of a hill were asked how steep they thought the hill was, in order to study biases of perception. "That was the tantalizing hook that I needed to study more of this . . . and Northeastern was the preeminent place because they made a very deliberate choice to bring in some of the leading network scientists in the field at the time."

As a graduate student in the network science program at Northeastern, Shugars said she was "really interested in the ways that things are connected to each other and the dynamics of how those connections work." While the program was described as being interdisciplinary, Shugars's work focused "particularly on political conversations and political reasoning, which I think we can agree is a complex system . . . and looking at the way individuals reason, but also the way people reason together about the matters that shape our lives."

27

The Complexity of Human Intelligence

We took a closer look at the human story of how we define intelligence. In the *Merriam-Webster Dictionary*, intelligence is defined as "the ability to learn or understand or to deal with new or trying situations." We felt that this definition and the cultural understanding of this word limits the holistic nature of true human intelligence. We were curious to learn more about the nuance, scope, and scale of human abilities, and we began to wonder what intelligence really meant.

Here are some of our favorite quotes that capture the complexity of true human intelligence:

> "The test of a first-rate intelligence is the ability to hold two opposed ideas in mind at the same time and still retain the ability to function."
> F. Scott Fitzgerald, *The Crack-Up*

> "Knowing others is intelligence.
> Knowing yourself is true wisdom.
> Mastering others is strength.
> Mastering yourself is true power."
> Lao Tzu, *Tao Te Ching*

> "Knowledge cuts up the world. Wisdom makes it whole."
> Brazilian proverb, cited by Christian Wiese in *The Way of the Meister*

> "The true sign of intelligence is not knowledge but imagination."
> Albert Einstein

> "Emotional intelligence is your ability to recognize and understand emotions in yourself and others, and your ability to use this awareness to manage your behavior and relationships."
> Travis Bradberry, *Emotional Intelligence 2.0*

28

Interconnectedness and Why Information Grows

For Episode 085 How Networks Learn, we talked with physicist, complexity scientist, and professor at Massachusetts Institute of Technology César Hidalgo. He is also the author of *Why Information Grows: The Evolution of Order, from Atoms to Economies*." In *Why Information Grows*, Hidalgo wrote: "Our ability to crystallize imagination sets us apart from other species, as it allows us to create in the fluidity of our minds and then embody our creations in the rigidity of our planet. . . . But crystallizing imagination is not easy. Embodying information in matter requires us to push our computational capacities to the limit, often beyond what a single individual could ever achieve. To beget complex forms of information, such as those that populate our modern society, we need to evolve complex forms of computation that involve networks of humans. Our society and economy, therefore, act as a distributed computer that accumulates the knowledge and know-how needed to produce the information that we crave."

This excerpt richly communicated Max Boisot and Bill McKelvey's Law of Requisite Complexity: "It takes complexity to beat complexity." Hidalgo calls for us to match the complexity of "embodying information" with complex, diverse, and knowledgeable networks. We need to come together and humble ourselves, because our individual knowledge is very limited, but we can maximize complexity and chaos through networks.

Hidalgo explained, "To learn more complex activities you need more social reinforcement." We make information grow when we leverage our creativity, imagination, and diversity. This begins with learning collectively. Having networks alone is not enough. We need meaningful interaction and collaboration in order for collective learning to take place.

To begin making meaningful interactions and collaborations we have to promote the right conditions for our networks to learn and adapt, which are incidentally the same conditions that enable life itself. Leading with emergence, fully engaged, while learning from experience and adapting along the way can help us foster open systems that enable growth. Also, allowing for emergence helps foster healthy and more dynamic relationships so that large amounts of knowledge can easily flow through and between networks. Safe-to-fail environments are also a great example of how to promote collective learning within a team environment.

On a more personal level, deep social bonds, trust, and community are critical. After talking with Hidalgo, we were reminded that the quality of our relationships really matter. He made a point to mention the people to whom he credits his life and his success. He also talked about the importance of building strong links or bonds with those closest to him, and he opened up about his concern for growing social isolation and loneliness in Western culture.

For us, our conversation revealed a deeper theme from *Why Information Grows*, which is that information has little meaning without relationship. When you strengthen connections, you create more possibilities, in every sense of the word.

29

I always tell my students and postdocs,
"the force is with us,"
and the reason for that is because
the world works as a network, we function within networks,
and there is no way we can understand this world
without using the tools of network science.

Albert-László Barabási
Episode 115 The Network Science of Success

30

A Conversation with Stephen Wolfram

We spoke with Stephen Wolfram, founder and chief executive officer of Wolfram Research, for Episode 101 A Journey of Computational Complexity. Wolfram was one of the founding fathers of the complexity science field and was the featured keynote speaker on the first night at the International Conference on Complex Systems 2018, hosted by the New England Complex Systems Institute. After speaking, Wolfram was trailed by a very eager crowd of complexity thinkers who lingered to ask him research-related questions and elaborate on his thought-provoking presentation. During this time, we anxiously waited in our podcast booth for him to come speak with us. After nearly two hours of patiently addressing his crowd's questions, Wolfram walked into our booth. We asked him if he was still up for an interview after such a long evening, and with no hesitation he sat down and put on his microphone headset.

Wolfram talked about his first introduction to complex systems when he was twelve years old. He discovered a book on statistical physics at this young age and his untethered curiosity would soon lead him to become a prodigy in the field of physics, writing his first scientific paper at the age of fifteen and graduating from the California Institute of Technology with a doctor of philosophy in theoretical physics by the age of twenty.

Throughout Wolfram's career, his fascination with physics and complex systems research would continue in parallel with his other passion for scientific computing. Eager to understand the origins of complexity in nature, he built himself a tool called cellular automata to study the behavior of simple computer programs. These computing experiments with cellular automata led him to some very surprising discoveries about the basic science of complexity. "I discovered what is now called 'emergence' through the surprising complexity of simple programs in the computational universe," he shared.

Wolfram explained in a grateful tone that his career has continued to iterate between intensive complexity research and scientific computing, and each of these two fields have led to discoveries in the other. He called himself lucky to be in his position, having passions which align in such a reinforcing way. His only disappointment was that he didn't engage more in unifying a complexity community that could focus intently on basic science research. After many attempts to bring together a very undefined and nebulous community of complexity researchers in the 1980s, he reached a crossroads in his professional life. Wolfram chose to shift his focus away from building community and more toward building the computing tools he needed to improve his research, and in 1988 he created the computational platform Mathematica.

Although Wolfram looked back on this decision with some regret, his career has been extremely successful and influential to the complexity science community. His widely-acclaimed, best-selling book, *A New Kind of Science*, rocked the scientific community with paradigm-shifting concepts and ideas, and his later work with Wolfram|Alpha has powered computing agents like Suri with artificial intelligence, a growing field closely connected with ideas in complex systems science.

The fact that Wolfram has any sort of regret about his influence on the complexity community demonstrates his undying passion for the field itself. His belief that the basic science of complexity should be foundational to everyone's educational journey is his truth and passion. Education has also played a large role in his life and career. He founded the Wolfram Summer School in 2003 and has been working to introduce young people to modern computational thinking.

What we found extremely profound in meeting and talking with Wolfram was how he not only helped to define the field of complexity, he embodied complexity itself. Reinforcing feedback loops continue to guide his work and he embraces the emergence in his research, allowing it to shape the tools he builds. He lives and breathes for this adaptation, which makes his work more useful and relevant for himself as well as others. There is a sense of harmony and coherence in his professional journey, a pattern that has emerged from many of the guests we have talked with at *Human Current*. Wolfram was inspirational. He shifted our mental models and widened our perceptions about what is possible in this incredible field.

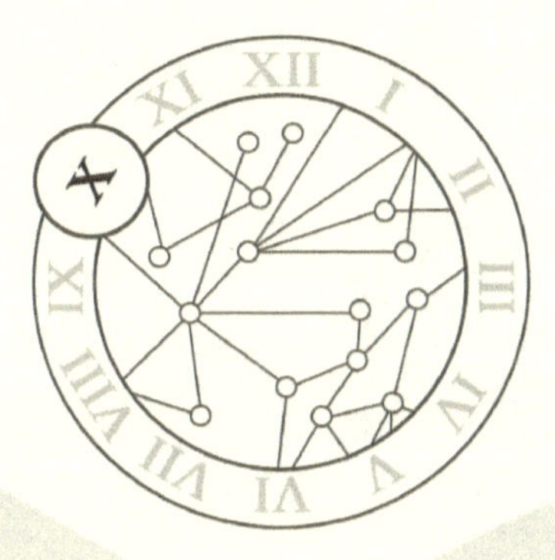

OCTOBER

1

Doug's Part in Revolutionizing the Printing Industry

Charles Ying, Richard Ying, and Douglas Drane standing in front of the building with "The Loft," where they started Atex.

With artificial intelligence and blockchain technology becoming more mainstream, we thought it was important to dedicate an entire series to these topics. As we reflected back on the series, we were in awe of how far technology has come in just a short time. We discussed the advances in technology with Doug, who reminded us of his own experiences with advancing technology in the 1970s.

Doug shared that while working for Computech he, along with his co-worker and friend Charlie Ying, had a vision for creating more affordable and efficient technology. When decision makers at Computech rejected their vision, they began to question why they were working for a company that did not share their views on marketing and company building, which led them to leave Computech. Doug then worked with Charlie and his brother Richard Ying to establish Atex, Inc. With very few resources (they each had a credit card with a $300 limit), they set out to make a system comparable to the mainframe IBM System/360 user interface that was faster and more affordable. They achieved their goal and essentially "revolutionized the printing industry from top to bottom," Doug revealed. Atex was later sold to Kodak for $77 million, and as Doug described, "The system took over the publishing world with Kodak adding color."

We have been inspired by how Doug and his business partners influenced the direction of technology and revolutionized the printing industry.

2

As the world becomes increasingly data driven
and increasingly more complex,
the companies that are going to survive
are the ones that enable their employees
with tools and empower them
to make good, data-driven decisions in real time.

Rico Andrade
Episode 049 Using Data and Design to Empower Teams

3

The Intersection of Diversity, Innovation, and Performance

We spoke with Paolo Gaudiano, an entrepreneur, teacher, and visionary who applies complexity thinking to all aspects of his work and life. We shared our conversation with him for Episode 048 The Intersection of Diversity, Innovation, and Performance. He is the president and chief technology officer at Icosystem, an organization that combines expertise in human behavior and decision-making with simulation and optimization techniques inspired by complexity science and biology to build predictive analytics solutions. Gaudiano said, "We would argue that every business is really a complex system, and when you take any action within your company, you get ripple effects that can have a profound, long-lasting impact on the business. Diversity is a perfect example of that."

Gaudiano, who is also a professor at The City College of New York, said, "The world is going crazy on the idea of big data. What people are doing is abdicating their knowledge and domain expertise and basically handing a bunch of data over to a mathematician that understands what to do with the data, but does not necessarily understand the business." He also observed that many of today's business tools are really intended to solve problems for complicated systems, and they don't necessarily work well for complex systems.

4

Just for the survival of the human race,
if they [artificial intelligences] don't have love
there's no logical reason why
they should support our survival. . . .
I think it is unethical to build something
that could be conscious and have subjective experience
that does not have the capacity to access love,
because having the capacity to access unconditional love
is what makes experience tolerable.

Julia Mossbridge
Episode 078 Loving AI

5

International Conference on Complex Systems 2018

At the International Conference on Complex Systems 2018 (ICCS), we spoke with Yaneer Bar-Yam and shared our conversation in Episode 092 ICCS 2018. He talked with us about the diversity of the presenters and attendees and shared his perspective on why the event was so important for the complexity science community and the world. "Thinking about what's happening in the world requires these new tools. One of the great and exciting applications of these tools is to the world as an entirety as well as many different parts of it, so that's why complexity science is important and this conference has really declared to the world that this is an opportunity, that we can engage in this new science to expand our understanding to new domains." Bar-Yam described that for the conference, "The idea is that we're at the beginning of an exploration of how we can understand in new ways many different systems."

6

Exploring Artificial Intelligence

We chatted with Melanie Mitchell for Episode 077 Exploring Artificial Intelligence. Mitchell's background is in mathematics and physics. She became curious about how intelligence emerges from a complex system after reading Douglas Hofstadter's book, *Gödel, Escher, Bach: An Eternal Golden Braid*. Hofstadter later became her doctoral advisor after she sought him out at the University of Michigan. Mitchell attributed her professional journey to Hofstadter and John Holland, a pioneer in the field of genetic algorithms. Holland was also an early founder of the Santa Fe Institute and encouraged Mitchell's involvement with the Institute.

Since her journey began, Mitchell has contributed to more than eighty scholarly papers in the fields of cognitive science, complex systems, and artificial intelligence (AI). She authored *An Introduction to Genetic Algorithms*, and *Complexity: A Guided Tour*, winner of the 2010 Phi Beta Kappa Book Award in Science. We were very excited for the opportunity to talk with Mitchell. Her history, career, and research have given her a unique understanding of the dynamics of systems, especially technological systems such as artificial intelligence and machine learning.

Mitchell is writing a book about AI that addresses questions like, "How close is AI to human abilities in many different areas, and what are the ethical, moral, and social issues that we really should be thinking about?" We asked Mitchell if unintended consequences might arise from increased use of AI and she replied with specific examples of how this technology can, in fact, amplify inherent biases. She said, "AI systems can be biased by the data that they learn from," for example, gender bias in natural language processing. She described how AI recommendation systems, like job advertisements, often pick up on inherent gender bias in the data they use to learn, which might result in people receiving different job recommendations based on their perceived gender.

Mitchell continued with another example about bias in facial recognition software, explaining that when these systems are primarily trained on people who are white, they end up performing poorly at identifying the facial features of non-white individuals. We asked her, how can we combat bias in AI systems when we still haven't figured out how to address it in society? She explained that she doesn't think anyone has an answer to this question yet, but the starting point is understanding what the biases are in AI systems. Understanding why a machine behaves a certain way is not easy because these systems do not explain themselves. Mitchell believes creating AI systems that can explain themselves is critical to understanding their behavior, biases, and vulnerabilities.

Mitchell's advice for navigating our relationship with AI as it becomes more popular was "to be very careful knowing when we can trust these systems." She believes we tend to trust algorithms more than we should, especially when we don't understand how they work. It seems we need to continually question the AI systems we build and interact with in order to work toward a better understanding of how they operate. Artificial intelligence is a powerful new tool that requires diligence and acute awareness of what is happening while it is happening. There is hope in building a better future with AI, as long as the farther we travel down this path, the more we practice our own self-awareness and conscious behavior.

7

The release of control in the right way unleashes something that you could never have planned for otherwise, and it allows a lot of opportunity for new insights and new execution to occur.

Rico Andrade
Episode 049 Using Data and Design to Empower Teams

8

The Myth of the Dysfunctional System

For Episode 040 The Myth of the Dysfunctional System we asked past guests Benjamin Taylor, Angela Montgomery, Oleg Konovalov, and Sonja Blignaut to share their thoughts on the following Jeff Lawrence quote from *The Practice of Adaptive Leadership*, by Ronald Heifetz, Alexander Grashow, and Martin Linsky: "There is no such thing as a dysfunctional organization, because every organization is perfectly aligned to achieve the results it gets."

Taylor responded, "You have to appreciate why something [a system] is the way it is and how it's learning to do that. And, you have to value what it's creating and achieving. . . . If you're looking at a system and you know what's wrong, take a deep breath, wait, and see what happens next; see what you learn before you act and I think you might go a lot farther."

Montgomery said, "We can say that an organization is dysfunctional when it works in a way that is not consistent with the goal it is supposed to pursue, so lack of clarity and understanding of the goal could easily produce dysfunctional organizational behavior, which means behavior that is not consistent with that goal. . . . What could be perceived as dysfunctional toward a stated goal can actually be extremely functional for a non-verbalized one."

Konovalov said, "Many organizations tend to lose their sharply defined goals or directions by rushing into something irrelevant, so wrongly-set goals are making them dysfunctional. As a result, they are losing their core functions and abilities. I call this 'goal perplexity.'"

Blignaut responded, "[This quote] perfectly describes how stuck we are in a mechanistic metaphor. We describe a system or a culture as being broken or needing fixing, as if it's a machine. . . . We tend to think about self-organization as a positive thing, so we talk about self-organizing teams in the Agile space, for example. But complex systems or human systems are *always* self-organizing, and they may not be organized in a way we think [they] should be. This is something that we see as dysfunction, but it is something that is serving part of that system."

9

People think these machines [AI]
will be able to solve all their problems
and the reality is, that's not the case.
You can use these technologies
and gain important capabilities,
but it's also important to frame these technologies
in the right way so you can make decisions
that will actually help you
rather than increase the risk of bigger failures.

Alfredo Morales

Episode 079 The Integration of Human and Artificial Intelligence

10

Cheers to 100 Episodes!

For *Human Current's* milestone Episode 100 Cheers to 100 Episodes, we gave cheers to all the wonderful people who have supported *Human Current's* mission, including Doug and the other co-founders, Stacy Hale and Lindsay Patterson. We also gave cheers to Yaneer Bar-Yam and the community of the New England Complex Systems Institute (NECSI) as we reminisced about our experience at the International Conference on Complex Systems 2018. We shared two interviews from the conference in this episode. One of our interviews was with Richard Cooper, a NECSI board member and a professor of international economics at Harvard University. Cooper said, "I'm an economist, and modern economies are extremely complex. It's a challenge for economists to simplify them, down to their basics, to the fundamental things that are going on."

The other interview we shared was of Emma Towlson, physics professor at Northeastern University and a researcher in the Center for Complex Network Research, with interest in the emerging field of network neuroscience. At the conference she gave a talk about how she was bringing complexity theory to high school students. "The world we live in is incredibly interconnected and you see that from this conference. This wouldn't exist if there wasn't such a demand for this kind of science. The world around us has changed in that way, but educational systems are still very linear in the way that things are taught. You're first taught this concept, and this concept, and this concept, and subjects are taught in isolation instead of acknowledging the interconnectedness between these things. And kids are very natural connectors and adapt to these ideas very easily at a young age, and as you get older they become harder to adapt to."

11

Applied Mathematics and the Evolution of Music

In Episode 097 Applied Mathematics and the Evolution of Music, we talked with Natalia Komarova, professor of physical sciences at the University of California, Irvine. She told us, "Evolution penetrates many different areas of science and society, and applied math is very versatile. It can help you study a lot of different interesting phenomena." She also described the simplest complex system as "a system where we have well-defined, simple rules, but the outcome is counterintuitive."

We asked Komarova about her International Conference on Complex Systems 2018 presentation, which explored the seemingly disparate topics of the spread of mutants and the evolution of popular music. She said, "My presentation had two parts: one was more biological application of my work and the other one was the fun application." For the first part of her presentation, the biological part, she "tried to formulate sort of a definition of a complex system that's the simplest." She said, "Biofilms are complex communities of bacteria. In these communities . . . we want to see how mutants develop because that may mean resistance to antibiotics. Everybody wants to know how to fight resistance to antibiotics, and so we would like to understand that."

Komarova described the second part of her presentation as "fun, and that was about the evolution of music," which she believed "was received by the audience much better because everybody wants to know about music!" She said that her research stemmed from hearing her daughter's music and thinking, as she put it, "Oh my god, I don't like this music! This is not what I used to listen to." Komarova went on to describe how she applied her expertise in the field of evolution to music. She and her research team reviewed music from the last thirty years and discovered "a way to extract, for example, the happiness index from a song" and a song's danceability. They concluded, "Music is becoming less happy. At the same time songs are becoming more danceable, which is not actually a contradiction. If you listen to modern songs they're very rhythmic. You want to dance to them but they're not as happy as they used to be thirty years ago. Another interesting [finding] is thirty years ago, sixty percent of songs were sung by males. Now that's reversed and most songs are actually sung by a female. So, songs are becoming more female."

She and her team also compared successful songs to all songs, defining "successful" as making it onto the top 100 charts. "Only about four percent of all songs made it onto the charts, and we wanted to look at the trends among those songs. Interestingly, we saw that these successful songs behave almost like a different species. They have different trends; for example, they are happier than regular songs. So, even though the overall trend is becoming less happy, the most successful songs are happier, they're more likely to be sung by females, they're also more relaxed and less aggressive."

12

For any business that can use blockchain
[to improve their processes] and is not using it now,
I think it's a race against time right now,
so educate yourself because
it's a matter of survival for your business.
Especially educate your leaders.

Mihaela Ulieru
Episode 120 The Social Impact of Intelligent Systems

13

Loving Artificial Intelligence

For Episode 078 Loving AI, we talked with author and systems thinker Julia Mossbridge. Mossbridge is the director of the Innovation Lab at the Institute of Noetic Sciences and founder of the Mossbridge Institute. She was also the principal investigator and team lead for the Developing Loving Intelligent General AI (Loving AI) project.

Mossbridge shared details of the Loving AI project, which "addresses how AI agents can communicate unconditional love to humans through conversations that adapt to the unique needs of each user while supporting integrative personal and relational development." The reason Mossbridge was involved in the project was, as she stated, "Not because I'm an AI expert; I'm a human being who sees the potential future of AI as being disastrous and sees this as a possible way to shift that." She also shared why she believes it's so important for AIs, specifically general artificial intelligence machines, to experience unconditional love. She said, "If we develop these [general artificial] intelligences, it's really important that they are loving, that they can access love. I know it's important for the future of humanity that they can access unconditional love."

Mossbridge also stated, "I don't care how many robots we have running around or how many artificially intelligent systems we've got, your experience of your life is always an inside job. And there's really no other way to fix it other than to be in contact with yourself and to love yourself."

14

The blockchain is really a link between all these ideas that have been happening in systems theory and complexity theory over the past decade and what is happening out there in the world at the moment. These networks are going to be extraordinarily disruptive, and I think they are going to take a lot of these ideas from complexity and bring them out into the world.

Joss Colchester
Episode 089 Nonlinear Systems and Technology with Complexity Labs

15

Health and Aging as Complex Networks

For Episode 124 Health and Aging as Complex Networks, we talked with professor and systems thinker Alan Cohen about his research on the aging process and holistic physiology. Cohen explained that controlled experiments continue to dominate the way biologists learn and think. He suggested, "This paradigm of complex systems thinking is what can guide us to understand when those controlled experiments are formative or not. The more complex the system, the more a lot of factors might be interacting, the more we have to doubt how generalizable our experimental results will be."

Cohen also shared some of the complex questions he is researching in his lab, such as, "Can we integrate biomarkers—like cholesterol, albumin, hemoglobin, glucose—together to get a good idea of what's happening with the aging process? And what happens if, instead of looking at them [biomarkers] one at a time, we try to integrate their signal and look at what might be happening in terms of underlying processes?" He said his research has demonstrated that people who have "high dysregulation levels" are at a greater risk of health complications. Cohen was excited to share that his research in holistic physiology is uncovering fundamental processes in the biology of aging. "I think we now have pretty good evidence that these complex systems dynamics are part of the aging process. They're probably not the whole story, but an important part."

We asked Cohen, if we do find a way to expand the human life, how does that benefit the greater whole of society and the planet? His response explored the many negative, unintended consequences that could arise from prolonging the human life, including an unsustainable population on a finite planet and severe social inequality.

16

If you are in the paradigm-shift business, the best indicator of good final outcomes is that people are really upset when the thing first arrives.

Stephen Wolfram
Episode 101 A Journey of Computational Complexity

17

Nonlinear Systems and Technology

In Episode 089 Nonlinear Systems and Technology with Complexity Labs, we talked with the founder of Complexity Labs, Joss Colchester. Colchester said, "One of the things that I like a lot about complexity theory is that it actually reopens a lot of science and math to philosophical and even spiritual ideas." Colchester described why he finds nonlinear dynamics and nonlinear systems fascinating, "A lot of things that sound quite technical, like feedback loops, chaos theory, the butterfly effect, attractors and bifurcation . . . I don't think they are taught to fit together properly."

He also discussed artificial intelligence (AI) and the challenges and opportunities it presents. He said, "It [AI] is less about machines. It's more about us and trying to understand ourselves and the fact that we now have these other entities, these automated systems, and that we have to actually differentiate ourselves with respect to that. We have to find a synergy to work together with those systems, and we have to move out of that space of doing routine mechanized activities and create a new function for ourselves."

Complexity Labs is now Systems Innovation, and Complexity Labs content can be found there at systemsinnovation.io.

18

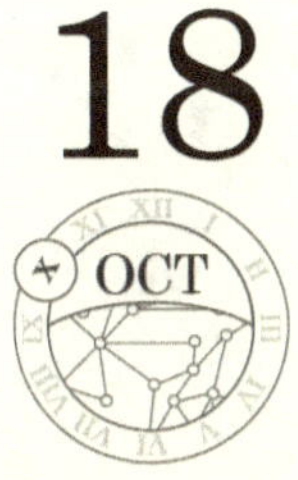

I don't devalue big data,
I do think it's important,
but if it's the only way and people
aren't digging into the human side,
we are missing half of the equation.

Spencer Gerrol
Episode 042 The Neuroscience Behind Your Vote

19

Getting these systems to be more explainable so that we can understand them and understand their biases and vulnerabilities is really important, and I think that's increasingly what a lot of the people in the AI research community are looking at.

Melanie Mitchell
Episode 077 Exploring Artificial Intelligence

20

The Social Impact of Intelligent Systems

For Episode 120 The Social Impact of Intelligent Systems, we interviewed Mihaela Ulieru. She is a scholar of distributed intelligent systems and founder and president of the IMPACT Institute for the Digital Economy. Ulieru said her true passion is "to analyze how technology is a provocation to human nature and how we coevolve with it."

Ulieru talked about the interplay between society and technology and its effects on our humanity. She shared many paradoxical examples of how technology, like artificial intelligence and blockchain, can help us transcend our limitations while also preying on them. "With blockchain, now we have a social network operating system with trust embedded in the network effect," said Ulieru. "When it comes to business there is so much value that the blockchain unleashes, value which is now caught in counterproductive processes." Ulieru also urged leaders to educate themselves on the ways blockchain can streamline their business, stating it's now "a matter of survival."

For all of us, she recommended, "Educate yourself on how data and information is misused and the advantages of decentralization in this regard . . . and be proactive rather than waiting for the government to give you help."

21

If you are solid in yourself
and your relationship to yourself,
you're much less likely to be in danger
of being taken up by sparkly-shiny-new
if it doesn't meet some need in yourself
that's valuable or worthy.

Julia Mossbridge
Episode 078 Loving AI

22

Answering Hard Questions

For Episode 095 Answering Hard Questions, we talked with Ryan Singer, head of strategy at Basecamp. We agreed with his statement, "The echo chamber is the enemy." Singer discussed design strategy, nonlinear dynamics, and interdisciplinary work. He also talked about the "fat tail" phenomenon, "A bunch of little things moving become a bigger thing moving, and then it becomes something that you can see at a large scale that is some kind of a coherent phenomenon, but all the little parts have to be moving in the same direction."

Singer gave the example of bestselling books, stating, "There's a way that the information about the book can get passed on, and as more and more people read it, the scale of the sales is blasting up because of this kind of activity that's blowing up. So, it's actually the connections between people that explain how a book can be a mega-seller compared to other books." He used this example to reflect on the growth of Basecamp and figure out "what we might do differently . . . it just triggers so many interesting thoughts and questions."

23

Complex systems science is not only the field that bridges across multiple disciplines, but also it gives you the framework to understand complex social phenomenon.

Alfredo Morales
Episode 079 The Integration of Human and Artificial Intelligence

24

Nonlinear Systems and Technology

In Episode 102 Probability and Adaptation, we shared a conversation between Gad Saad and Nassim Nicholas Taleb about their joint presentation at the International Conference on Complex Systems 2018. Saad is an evolutionary behavioral scientist and marketing professor at the John Molson School of Business, Concordia University, Montreal, and Taleb is a best-selling author, probability researcher, and former trader.

Saad and Taleb defined "the use of mechanistic techniques in statistics" as a problem, stating, "There is no such thing as a standard statistics." They pointed out, "There is a crisis in science where you have researchers coming up saying, 'You're doing this wrong,' when their science is a random variable. It's not science." Taleb explained, "What people don't know is that the 'p' value is also a random variable, just like correlation is a random value."

25

I think the problem that I have with best practices
is the word "best," because it implies
that there is no better way to do things.
But as we learn, and as we apply our outside-of-work
experiences to work and vice versa,
we find better ways to do things
through the way that we evolve
with the systems that we are a part of.

Stacy Hale
Episode 012 Systems Thinkers Share Their Stories

26

Learning How to Thrive in Complexity

For Episode 125 Learning How to Thrive in Complexity, we talked with writer, researcher, and coach Jennifer Garvey Berger about her passion for spreading ideas about complexity, leadership, adult development, and change "so they can do the good that the world needs them to do." She explained, "There are ideas about the world that are unbearably useful, and yet they are often locked up away from the people who could best use them." Berger's work and research are driven by her desire to get ideas about complexity out into the world. She expressed to us that there is joy and beauty in complexity, describing it as "a fundamental energy in the world." Her recommended approach to the complexity of our world is that we should try to understand the rules and learn to play inside them, rather than try to contain and control complexity.

Berger shared insights from her book, *Unlocking Leadership Mindtraps: How to Thrive in Complexity*, including the five mindtraps she believes are holding us back. These are: our desire for simple stories, our sense that we are right, our desire to get along with others in our group, our fixation with control, and our constant quest to protect and defend our egos. She said, "If we can get a handle on some of these mind traps, that redeploys our energy and lets us work with the system instead of continually fighting against it."

27

The daunting issues that face humanity and our planet aren't going to be solved through technology. They just aren't. Technology already exists. The problem is in consciousness.

Rollin McCraty
Episode 073 Coherence and the Heart's Intuitive Guidance

28

The Integration of Human and Artificial Intelligence

For Episode 079 The Integration of Human and Artificial Intelligence, we talked with assistant professor at the New England Complex Systems Institute (NECSI) Alfredo Morales. Morales works in the areas of complex systems, artificial intelligence (AI), data science, and human behavior to develop methods and insights that help solve complex societal problems. He described complex systems science as "the bridge between both the quantitative side and the social side," and said, "Complex systems science is not only the field that bridges across multiple disciples, but also it gives you the framework to understand complex social phenomenon."

He shared details from the AI and Beyond program at NECSI, a five day certificate program where he presented alongside Stephen Wolfram, Iyad Rahwan, and Yaneer Bar-Yam. "In these programs [at NECSI], we train people in understanding the fundamentals of complex systems as well as the tools that are available for them to cope with it, and AI is a very important piece," said Morales. He also discussed some unintended consequences that could arise from AI and how complexity science can help us integrate AI systems more effectively. Morales shared, "My contribution in the [AI and Beyond] program is to explain what AI is about so it stops being a black box that just yields good results . . . so you really understand what the machine is doing and why it is behaving the way it does. And I link this with complexity thinking, so you can also frame why the machine is able to learn and what are the limitations as well."

29

There are so many things in our lives that are joyful, beautiful, glorious because they are complex, and I am completely uninterested in "managing" that away.

Jennifer Garvey Berger
Episode 125 Learning How to Thrive in Complexity

30

What a One-Man Orchestra Can Teach Us About Complex Systems

In a complex adaptive system, feedback shapes how change happens. Feedback loops form when there is mutual causal interaction (x affects y, and y affects x). Since feedback loops and Andrew Bird are two of our favorite phenomena, we were pleased to uncover a video of this concept set to song in a musical *TED Talk*, "Andrew Bird's One-Man Orchestra of the Imagination."

At first, Bird described feedback loops in terms of music and sound, "When the microphone gets too close to its sound source, it gets in a self-destructive loop that creates a very unpleasant sound." Then with a few musical examples to carry the metaphor, he related this to effects that occur in complex systems. "There seems to be a rule in nature that if you get too close to where you came from, it gets ugly." Bird's album, *Break It Yourself*, is an ode to creative destruction. Hewn from a series of elegantly constructed looping phrases, the songs tell of disrupting the downward spiral of heartbreak with acceptance and "reckless curiosity."

Like anything in complexity theory, a feedback loop is not inherently good or bad. As we understand the concept in complexity, positive feedback loops create attractors, or self-reinforcing associations among co-evolving agents, so that the system produces more of the same behavior. Negative feedback loops disrupt the pattern and produce novelty, resulting in innovation in the best cases, or putting the "brakes" on what could, in the worst case, become a destructive spiral.

Bird touched on the phenomenon of self-consciousness as just this sort of flirtation with the insight of one's own undoing. He sang, "And the eye that eyes itself is your eye; and the ear that hears itself is near." We found this much to the tune of Alan Watts, who referred to anxiety and mental feedback loops as "the quaking mess."

31

In complexity, a simple story is really problematic because while they travel well, they travel easily, and they hold a lot of our culture, they also really constrain us from seeing new possibilities and from seeing one another fully.

Jennifer Garvey Berger
Episode 125 Learning How to Thrive in Complexity

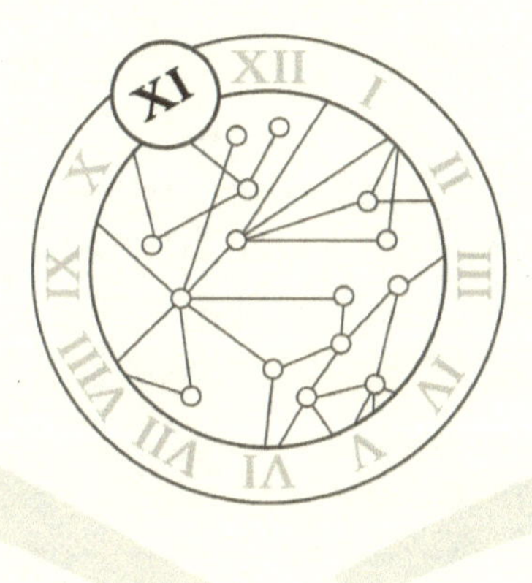

NOVEMBER

1

Team Dynamics and High-Performing Teams

Doug taught us that high levels of complexity need to be matched with equally high levels of complexity. He would quote Max Boisot and Bill McKelvey's Law of Requisite Complexity: "It takes complexity to beat complexity." To Doug, organizations with hierarchies, silos, and top-down leadership were structurally ill-equipped to handle complexity because of their slow communication channels and information bottlenecks.

Doug believed in the power of high-performing teams and team leaders who were responsible for systems-level understanding. He would often remind us that leaders should model self-responsibility and other critical problem-solving skills, so that the rest of the team could learn from their behavior. Doug saw leaders as mentors, role-models, and facilitators who guide and support the team's overall mission with point-to-point communication channels and transparency. Layering talent and facilitating team mentorship at all levels is how Doug thought flat organizations should be structured to manage complex problems.

2

The New Reputation Economy

We live in a reputation economy where intangible assets like trust make up a brand's market value. In Episode 016 The New Reputation Economy, we asked "reputation whisperer" Bonnie Caver of Reputation Lighthouse about designing a solid ecosystem for the reputation of a brand. She explained, "About 60 percent of purchasing decisions are based on the company, 40 percent are based on the actual product or service." She also discussed her seven dimensions of reputation: products and services, citizenship (engagement in the community), innovation, workplace, governance, leadership, and performance. As co-host Stacy Hale stated, "A company's brand is an ecosystem co-evolving with its employees, stakeholders, and customers."

3

A lot of the failure that you see in economic systems, political systems, social systems, corporations, and so on, comes from the fact that the people who are in charge of making decisions don't understand the complexity of the system they are dealing with and the effects their decisions have in a complex environment.

Alfredo Morales
Episode 079 The Integration of Human and Artificial Intelligence

4

Connections, Not Quick Fixes

There are many instances in our society when we resort to reductive, reactive responses to social problems. As Daniel Taber said in Episode 035 Using Systems Science to Improve Community Health, "Unfortunately, we don't like complexity in our society; we're always looking for the quick fix." If someone commits a crime or acts out, we punish them. If educators struggle to evaluate school performance, we implement standardized testing. We often reach for a counteractive policy or punishment, like a Band-Aid for a wound. Unfortunately, these quick fixes come with many unintended consequences, which sometimes exacerbate the problem they were designed to solve.

These consequences happen because social problems lie within many interrelated systems, so there is no isolated or linear way to approach them. They are *social* problems after all, meaning they involve connection and people, and they depend on networks to reach a solution. Connection creates opportunity. When it comes to social systems, it appears the more we reach for quick fixes, the more off-track the system and the agents within it become.

5

Tolerance in Social Systems

Connection breaks down barriers within a network, giving people access to more relationships, interactions, feedback, and growth. Strong and weak ties within a network help it evolve and grow.

We know that connections strengthen social networks, and yet we constantly resort to rejection or isolation as a way to "resolve" social issues. When using this tactic, we create a cycle of deflection, where the person or "problem" is consistently pushed out of one system and into another. This mindset, which is geared toward eliminating the problem, is in itself a problem.

As an example, think about the "zero tolerance" policy in schools. The National Clearinghouse on Supportive School Discipline describes zero tolerance as "school discipline policies and practices that mandate predetermined consequences, typically severe, punitive, and exclusionary (e.g., out of school suspension and expulsion), in response to specific types of student misbehavior—regardless of the context or rationale for the behavior." The majority of young people who are affected by zero tolerance policies are often labeled as "problems" because of their daily life struggles, such as poverty, neglect, and racism. Ironically, these students require the most support and tolerance for error. They are not "problems." They could be disconnected, lost, and frustrated. They need a support system to help them navigate their learning journey.

Sadly, many of the students who are punished according to the harsh guidelines of zero tolerance policies are expelled from school or placed in juvenile detention. Once they are cast out of the educational system into the criminal system, they fall behind in their studies and struggle to graduate. They become trapped in an ongoing cycle, which inhibits learning and growth. How can we expect any growth to occur in a system which leaves no room for mistakes?

Helen Keller said in *To Love this Life: Quotations*, "The highest result of education is tolerance." When dealing with social systems, we must practice tolerance. We can't expect to find resolutions to our social problems by holding on to prejudice. No policy or punishment will "fix" human behavior, but well-designed systems can provide support for people to reach their potential. Then they can *choose* to take the opportunity.

6

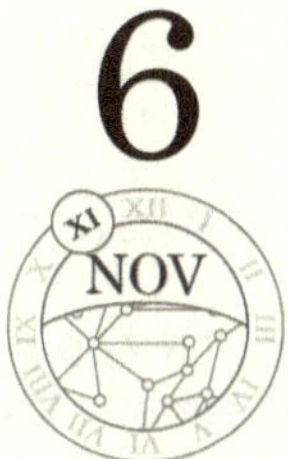

Paradigm shifts very quickly get framed
in a from-this-to-that, either-or type way.
And when you take the longer arc of the human endeavor
and you look back, then you realize
that every time we did that [paradigm shift]
we threw out the baby with the bathwater.
We didn't ask ourselves enough,
what is valid in what we're currently doing
that we actually cherish and need in the future?

Daniel Wahl
Episode 122 Design-Led System Transformation

7

Democracy Is On Us

When we start to think about the complexity of politics in the United States, it is easy to feel overwhelmed. After all, most of our exposure to politics began when we were children witnessing emotionally charged conversations between adults. They volleyed back and forth and once the debate was over, nobody looked happy. We are given the impression that politics is an intimidating topic right from the get go. Perhaps this should be expected when dealing with opposing ideologies.

The opposition-driven framework, which is integral to the United States government, has pressured citizens into believing that there are only two roads to lead us forward: one going left or one going right. So, we have to pick a side or get left behind in the dust. This blue or red, right or left, Republican or Democrat, wrong or right two-party system is increasingly creating more division in the United States. This makes us wonder why a three (or four) party system hasn't been truly accepted.

For Episode 041 Compassion in Chaos, we talked with columnist, lobbyist, and Republican strategist John Feehery on the complexity of politics in the United States. Feehery described his political outlook and shared how and why democracy requires a lot of personal accountability from all Americans. He stated, "I do think that democracy is all on us, and there are ways to have your voice heard despite all the complexity. . . . We have to actively participate in it and be hopeful that a better future is ahead of us." Feehery also stated, "If they [American people] don't get involved themselves, they're going to leave it up to the ideologues and extremists and that's what I think has often happened in our democracy, where debates are overtaken by ideologues and not by people who particularly want to make this country better and solve problems."

Feehery recommended sharing our stories as one way that citizens can be heard. "When people are passionate, even if they don't have money, even if they don't have great political connections, but if they have a great story and a great narrative, they can make a huge difference."

8

*Even if people are uninformed,
voting is important because
high voter turnout contributes to stability,
while low voter turnout will lead to instability
and negative representation.*

Alex Siegenfeld
Episode 114 Your Vote Brings Democratic Stability

9

The Emerging Narrative Shift in American Politics

We believe the ongoing narrative of bipartisan opposition is changing in America. There is a new narrative emerging which is unified by this one question, "Why are we fighting each other instead of working together to solve the complex problems facing our nation?"

Many Americans are beginning to recognize how paralyzing bipartisan opposition can be; they also realize that most people agree on and hope for the same things, like financial stability, economic prosperity, national security, and equality of opportunity. They know that we can't continue forward with the same mental models that created this bickering mess of a political system, especially since these mental models are no longer serving us or our country.

This group of dissatisfied citizenship is breaking the mold. They are seeking alternatives to their bipartisan government. According to Gallup in "American Identification as Independents Back Up in 2017," 42 percent of Americans identified politically as independents in 2017, up from 39 percent in 2016. This climbing number of individuals identifying as independent is an unintended consequence of the political system's rigidity—and untrustworthiness. Most politicians are viewed as superficial figureheads who only talk in circles, leaving no room for average citizens to discover the complexity and nuance hidden under all the talking points.

Fortunately, people like Joe McGovern are cutting straight through the talking points to get to the "meat," the complexity, the rawness that lives within politics. In an article for *The Independent Voter Network*, McGovern said, "Most things of importance are complex and the world is filled with nuance. The way we talk politics in the United States, however, doesn't allow for much of this complexity and nuance. We oversimplify things. We generalize. We hyperbolize."

McGovern grew so tired of the extreme political division in America that he decided to film a documentary, *The Other Side: A Liberal Democrat*

Explores Conservative America, in order to find a better way to talk about politics in America. The story of his journey is truly inspiring, and McGovern exemplifies a growing number of people in America who are driven to shift the polarizing narratives that currently dominate American politics.

10

The Neuroscience Behind Your Vote

We talked with Spencer Gerrol, human behavior specialist and founder of SPARK Neuro, for Episode 042 The Neuroscience Behind Your Vote. He described the role of emotions in decision-making: "Every decision is influenced by emotion, emotion plays a major factor in every decision you make in life, and without emotion you can't even decide. . . . You cannot, as a human being, make a decision without emotion. It is impossible."

This insight about the nature of human decision-making led him to his work's purpose. "If we can get deeper insights as to why people are making decisions and how emotion has an influence on that, we can lead innovation in a much more real way." His research to gather biofeedback data from both Trump and Clinton supporters demonstrated his ability to predict people's voting preferences by observing their emotional arousal. The study was discussed in a *Washington Post* article, "Whom Are You Voting For? This Guy Can Read Your Mind."

Gerrol said, "It's very valuable for people to understand the value of their subconscious. . . . We think we are rational beings, and if we go on assuming that and not really understanding how we make decisions in our everyday lives, then we actually act even more irrationally." We thought it was fitting, given Gerrol's research, that he quoted Spiderman's uncle, "With great power comes great responsibility."

11

Your Vote Brings Democratic Stability

In Episode 114 Your Vote Brings Democratic Stability, we talked with Yaneer Bar-Yam, president of the New England Complex Systems Institute, and Alexander Siegenfeld, a doctoral physics student at the Massachusetts Institute of Technology, about their collaborative research paper, *Negative Representation and Instability in Democratic Elections*. Siegenfeld said, "We were interested in, what does representation mean? What are the general properties of elections? Why are we in this current situation?" He explained that over the course of the research, "We discovered that the core problem is something we term 'democratic instability.'"

The researchers discussed why the current political climate in the United States is so polarized and unstable. Bar-Yam described an interesting observation, "When you have polarized voters you don't necessarily get polarized elected officials. In particular, if more people vote, it tends to reduce the polarization of elected officials." Bar-Yam and Siegenfeld explored how low voter turnout leads to negative representation and further instability. Siegenfeld claimed, "Whenever you have an unstable election, there will always be some opinions that are negatively represented." Siegenfeld used the metaphor, "A boulder sitting in the bottom of a valley is stable, while a boulder perched at the top of a hill is unstable. Where do elections sit?" They shared the importance of increasing voter turn-out and weakening the grip of the two-party system through methods like ranked-choice voting in order to achieve a more stable democracy.

12

Post-Election Awareness

The morning after an election is a disappointing time for at least half of the people who ran for office, supported legislation, and voted. The only thing to do is move forward. And to do that, as Dave Snowden said to us in Episode 044 Making Sense of Human Systems, "We have to start from where things are, not from where we would have liked them to be, or think they should have remained."

Awareness helps us figure out what happens next. By focusing on the present moment, we can recognize when we're in an echo chamber or feedback loop which is only reinforcing ideas similar to our own. We can push ourselves out of our comfort zone and make a conscious effort to see things around us differently. Small changes within our everyday lives will lead to lasting impacts. Every interaction is a chance for us to practice our awareness. We can ask ourselves questions like: What assumptions am I making about this person or people? Am I assuming good intent, or fearing the worst from people? How am I being or not being empathetic? Do I really understand this person? Am I listening to understand, or just to reply?

There is very little in this world we can actually control beyond our own actions, but we have to practice mindfulness every day in order to earn even that level of control. Pause. Breathe. And think before you respond. As Bohdi Sanders wrote in *Martial Arts Wisdom*, "When you react, you let others control you. When you respond, you are in control."

13

Conscious Voting Through Political Curiosity

Leading up to the 2016 elections in the United States, we explored our political mental models in order to question our own assumptions and ideologies. "Not everything aligns with what we may believe," said *Human Current* co-host Angie Cross. We discussed presidential power and helpful resources for how to increase awareness of the complexity underneath most political talking points. Here is an excerpt from Episode 043 Conscious Voting Through Political Curiosity, as co-hosts Haley Campbell-Gross and Cross share and discuss comments by guests John Feehery, Republican strategist and columnist, and Spencer Gerrol, founder of SPARK Neuro.

Campbell-Gross: We shared a quote pulled from an article by Eric Loomis with both interviewees, "It's Not Presidents but Pressure Groups Who Lead US Politics" [*Aeon* magazine]. He stated: "Power in the US is unusually decentralized for a strong nation. The fact that there are so many levers to that power should undermine narratives of presidential leadership; alas, such complexity would not help television ratings."

Feehery: Well, I do think that power can be diffused and has been diffused, but I also think that the executive branch outside of the president has tremendous power. The unelected bureaucrats in this country have a lot more power over our daily lives than most of us recognize or that we really appreciate. Sometimes that diffusion of power means that the unelected bureaucrats have more power than ever, and I think that any citizen should be worried about that.

Gerrol: We can debate all day long about whether or not the president him or herself has an effect directly on policy. And whatever you net out of that argument, one thing that we can say for sure, we know that your feelings of pride in your country, your sense of having that role model that you and your children can look up to, your sense of identity and being a citizen, are affected by who's the leader in charge. That's no different when you look up to a role model who is a teacher, a coach, a parent, or the CEO of your company. That figurehead, regardless of their impact on the decisions, is important emotionally. That is as important [as policy], and something that we can't neglect.

Campbell-Gross: There are a lot of levers involved in power in government, but the president is more than that. He is an agent in the system, and we react emotionally to that position. And, that in itself is powerful.

Cross: In that quote and even in the responses [from our guests], we use that term "power," and it makes me wonder, does power imply order in the system? And if not, perhaps it's safe to say that there's chaos in the US political arena with actual hidden deterministic order.

Campbell-Gross: I would say there is definitely, in my opinion, some chaos in the US political arena. . . . Where is the self-responsibility, because human motivation can be many different things. We could be motivated by making the environment better, supporting our country, family, there are many things beyond money that motivate people. . . . I know emotion can be motivating and like Spencer said in our interview with him, you can't make decisions without emotion, it's impossible. But we need to realize that anger can hold us back from thinking through things clearly and from having the important conversations that really move us forward.

14

Post-Election Reflection

The results of the 2016 presidential election in the United States shocked many people. Some people were saddened, some angered, and some felt they prevailed and their voices were heard. All these emotional responses are common after any major election. However, we wanted to explore the possibility of a different response, one of curiosity. One that was filled with wonder and driven to try to understand the events leading up to the election. Because it is in the space of curiosity that answers are found.

Sadly, we often live in isolation. Even with the use of technology, we isolate from ideas that differ from our own, and this is very dangerous. As Trent Lapinski wrote in an article for *Medium* summarizing the 2016 presidential election, "This is the problem with America today, the technology that was supposed to bring us together actually isolated us into echo chambers and drove us farther apart."

Many of us aren't comfortable enough to get outside of our comfort zone, or to question the news and information that is constantly pushed to us. We aren't listening to each other. We aren't trying to understand points of view outside of our own. We aren't inquisitive enough.

Curiosity is healthy and empowering. We can learn so much, not just about the other side, but about ourselves. We can get curious about the past, and look back and question why Donald Trump was elected president—an act of reflection that we believe is necessary and important.

To begin our quest to remain open-minded, we can use our emotions to propel us forward and motivate us. If we channel our anger, fear, frustration, or confusion into wonder, we can redirect what might be holding us back. We can ask ourselves reflective questions, such as: What important signs and events throughout our society and around the world did we miss? What did we see and know about, yet dismiss as unimportant? Why? Did we make false assumptions? If we would have asked ourselves these questions a month before the 2016 election, would we have felt more prepared to hear the election results? Would the election results have been different?

15

Decisions that we make can be linked
to effects in other parts of the world
and what other people are doing
can affect us in ways that are
surprising, unanticipated, and, in general,
part of the unanticipated consequences
of actions that are being taken.
So, there is a tremendous challenge
to deal with the complexity in the world today,
and we are trying to explain to people
how to think about what is going on in the world
and how to begin to build systems that are going to be able
to really respond successfully in the context of
the high complexity of the world that we have.

Yaneer Bar-Yam
Episode 050 Complexity Science Is Everyone's Science

16

Open-Minded Reflections

Questioning the past is useful for understanding patterns and causality because the emergent properties which arise from complex systems don't become apparent to us until later, in hindsight. Daniel Wahl told us in Episode 122 Design-Led System Transformation, "In questioning, there is a regeneration happening. It's inviting you to look at the status quo and double-check your experience and relationship with it." These moments of reflective questioning can also release us from strong biases and open us up to the truth—even truths we may not like.

Questioning the past helps us practice resilience, so we can become more equipped to manage uncertainty as we move forward. We can learn how to approach uncertainty with a new, more open attitude. It's healthy for us to work through the past, but we should not live in the past nor dwell on it for too long. After practicing open-minded curiosity and working through our emotions and uncertainty, we can practice mindfulness and awareness of the present.

17

We need a different kind of leadership that understands tipping points.

Tanuja Prasad
Episode 118 Practical Applications for Interconnected, Nonlinear Systems

18

Certainty Merchants

We spoke with Diego Espinosa, the founder of Sistema Research, for Episode 007 Certainty Merchants. Espinosa talked about the way information travels in ant colonies, financial meltdowns, and human social networks. In a complex world where we are bombarded with false signals about certainty, we wondered, "How can we regain resilience?" Espinosa urged us and our listeners to "take back control of your personal relationship with uncertainty."

Climate change, inequality, the rising cost of college tuition; these are all complex problems. Out of touch with the rhythms and interdependencies of our natural environment, we look to statistics and experts to help us make major decisions. Are we missing something? We learned from this former money manager how our addiction to certainty over the last three decades has created a whole industry of specialists who make money using statistics to sell the promise of certainty and security—and it's making us more vulnerable. "Our personal relationship with uncertainty is fundamental to being human, yet over the last thirty years we've begun outsourcing it to other people," said Espinosa. He made "a very personal request" to our listeners "to re-establish and re-engage with uncertainty in their lives." He said, "Think about the fact that the future is uncertain. How do you feel about that? How does it make you feel?" He added, "You have a relationship to those big questions."

What happens when we stop listening to the certainty merchants? Can we leverage our strongest social bonds to regain resilience in an uncertain universe? What can the complex patterns of the natural world teach us about ourselves? Espinosa said, "I think if we see ourselves as a little bit less special, a little bit more part of nature, that we'll naturally start to think about things in a complex systems way."

19

From Venezuela to a Science of Collapse

For Episode 093 From Venezuela to a Science of Collapse, we interviewed Ricardo Hausmann, professor at the Kennedy School of Government and director of the Center for International Development, both at Harvard University. From him we learned about the scale of the current economic collapse in Venezuela, systemic resilience, and the invisible hand. He said, "Complexity techniques, network science, and so on allowed me to analyze the economy without destroying [relevant] information, and then I did a bunch of cool things with it. . . . We analyzed how this economic complexity grows. Is it important? Is it predictive? What makes its growth difficult? How to overcome it? And so on."

Hausmann described his international perspective on immigration. "I find that in many developing countries, immigration policies are incredibly tough, especially toward the higher educated." Hausmann claimed that this is because "you have all of these guilds and professional associations and so on that fear competition from foreigners, that protect their markets from foreign competition." He described how entry is restricted and that his research suggests "that is extremely damaging to a country."

20

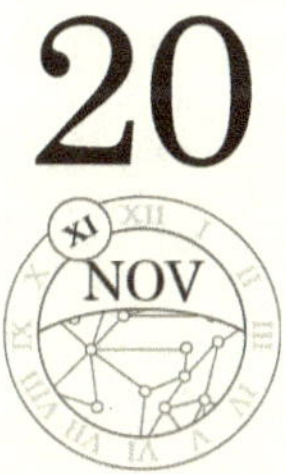

How Networks Learn

For Episode 085 How Networks Learn, we talked with physicist and professor at Massachusetts Institute of Technology César Hidalgo about his interest in the physics of networks and complex systems science. Hidalgo said, "The way that we think about entropy in physics and information theory come from different traditions and sometimes that causes a little bit of confusion, but at the end of the day it's the number of different ways in which you can arrange something." He said, "Thinking about complexity is important because people have a tendency to jump into micro explanations for macro phenomenon."

Hidalgo also shared why he believes these fields are so important and discussed how economic development is linked to making networks more knowledgeable. "I think complex systems give you not only some practical tools to think about the world, but also some humbleness because you have to understand that your knowledge and understanding of how the systems work is always very limited, and that humbleness gives you a different attitude and perspective and gives you some peace." He observed, "Everybody fails, but not everyone learns from their failures."

21

Networks and Interdisciplinary Research

For Episode 106 Networks and Interdisciplinary Research, we talked with Jose Mateos, a physicist and network researcher at the Universidad Nacional Autónoma de México. Mateos discussed how he started his studies in physics and then in the 1990s expanded his interests into chaos and complex systems, which was a new field at that time. Mateos has been involved with various conferences on complex systems and said what makes the events so unique and important for solving real-world problems is that complexity provides a common language for people from different disciplines. He said, "You can really communicate in an honest and deep way with these different disciplines, and we have the same common goal, for instance, understanding society, understanding the dynamics of a big city, and many of the real problems in the world are complex systems."

Mateos said approaching problems with just one lens—such as physics, mathematics, computer science, or social science—is a reductionist approach. He shared the example of migration, which he described as neither "a system of health or poverty, it has different viewpoints but from the viewpoint of complexity, you can grab the system and the essence. . . . So, it is a challenge, but it is fascinating." He claimed, "There's much more needed to really tackle the real complex problems that we have. . . . We have to make a team between complexity scientists and politicians to work together to solve problems."

22

The Path to Economic Growth

For Episode 067 The Path to Economic Growth According to Complexity Science, we interviewed the founding president of the New England Complex Systems Institute (NECSI), Yaneer Bar-Yam, who discussed a published NECSI research paper, "Preliminary Steps Toward a Universal Economic Dynamics for Monetary and Fiscal Policy." Bar-Yam, along with researchers Jean Langlois-Meurinne, Mari Kawakatsu, and Rodolfo Garcia, wrote this paper to explore why it took such a long time for our economy to achieve economic growth after the 2008 financial crisis.

The researchers asked a series of questions to direct their research: What are the issues confronting economic activity? Why isn't growth happening the way we would like it to? What does the economy look like as a whole? What things are most important in regulating the economy?

These questions led them to conduct a deep analysis of historical economic activity that uncovered an important path-dependence pattern. Bar-Yam explained, "Every few years the economy runs into the boundary of having too much investment, and then, there is a recession. In order to repair the economy, the Federal Reserve reduces interest rates because that's what it knows to do." He went on to explain that this regulatory method, which has been used about every ten years since 1980, "cannot fix the underlying dynamics of running into the wall." He used a metaphor to describe the way the government has been regulating the economy. "It's like driving a car with only the accelerator and brakes and not the steering wheel."

The major concern is we don't know if this method of adjusting interest rates will be effective in the next financial crisis. Bar-Yam said we need to rebalance economic activity before it's too late. But, what does it mean to rebalance economic activity? Bar-Yam and fellow NECSI researchers designed a system archetype diagram to analyze how money flows through the economy in two loops: the wages and consumption loop, and the investment and returns loop. They showed that both investment and purchasing power are needed to increase economic activity.

NECSI's findings also revealed that there is an imbalance in money flows, with too much money in the investment loop and very little demand coming from consumers, who have accumulated significant debt in recent decades. Bar-Yam and his fellow researchers called for an increase in money flow to the labor consumption loop. This means wealth redistribution, not tax cuts, is needed to improve the United States economy.

Bar-Yam elaborated further about how to achieve wealth redistribution through actions that would deal with consumer debt directly, including making changes to minimum wage, adjusting entitlement programs, or re-evaluating tax programs—all of which affect consumers' ability to participate in the economy. NECSI's findings demonstrate that addressing income equality is not just a humanitarian concern. It is aligned with the desire to make the economy work better overall, for everyone.

23

To let ourselves wonder at ourselves
is not just an intellectually pleasurable thing,
it's also politically important.

Simon DeDeo
Episode 108 An Exploration of Social Minds

24

Macroeconomics and Financial Crises

For Episode 096 Macroeconomics and Financial Crises, we spoke with Harvard University Kennedy School of Government professor Carmen Reinhart. Reinhart talked with us about financial fragility and economic booms and busts. She also explained the profound complexity of financial crises and described them as "black swan events" that affect many systems connected to human well-being. Reinhart said, "Uncertainty or lack of credibility is fundamental in crises during financial panics."

We asked Reinhart how she sees the connection between her work and the complex systems community. She responded, "Crises themselves are very complex. They are multifaceted, so I think that there are a lot of fields of research and fields of study that are outside the more narrowly defined realm of economics that are relevant and have a role in enhancing our understanding of these traumatic events."

25

When is Too Much Order a Bad Thing?

How much stock can we put in certainty? What causes financial panics? Most people blame politics, bad statistics or greed. It turns out it's more complex than that—but it's not all that complicated.

In Episode 007 Certainty Merchants, we talked with Diego Espinosa about how our quest for certainty has taken our market system to an unnatural place of positive feedback loops and financial inequality. We have unwittingly created a marketplace of "certainty merchants" who use the story of statistics to save us from our own anxieties, to the tune of bubbles and financial panics.

Espinosa pointed out that feedback loops are the basic building blocks of order created by complexity. "The second law of thermodynamics says that things tend toward entropy, and in self-organizing systems they tend away from entropy and toward order. A feedback loop is another way to describe self-reinforcing behavior. So it's organic to complex systems to have those types of patterns."

He reflected, "The thing I find so interesting about inequality is that it emerges from a system of competition. All of a sudden we had a very high degree of inequality that emerged. How did that happen? There are all kinds of explanations and some of them have moral qualities to them, that people are greedy, bankers are greedy. I don't dispute that, I just don't think it's the fundamental cause of inequality. I think the fundamental cause is that you have a system that tends toward fairness for a variety of reasons, but can spark a feedback loop that tends away from fairness if certain things disappear from the system. In our economic system we've gone from relatively stable . . . into a positive feedback loop that goes to more inequality over time."

We wondered how we can put the brakes on this inequality. Espinosa said, "The idea behind my article is that when feedback loops get engaged, there are natural brakes that exist in the system. If it didn't have these natural brakes, it wouldn't exist, because the inequality would get so bad that people would opt out of the system, have votes or revolutions for other systems." He described three fundamental brakes that kick in: competition, political processes, and financial crises.

Espinosa reminded us that greed is natural, part of human nature, and that sometimes it produces diversity in a network, and sometimes it produces homogeneity. Espinosa said, "We tend to think of financial panics as really bad things because they hurt people, and I understand that and sympathize with it. However, it is also true that if we don't have them, inequality will build up and build up, and that creates a basic sense of unfairness in society that is destructive to it. Personally, I'd rather see the pain of financial panics than the far more destructive power of a complete erosion of fairness and what that could produce."

26

If we understood emergence,
we would understand our work better
because so many of us,
myself included until about eight years ago,
were totally focused on changing large systems
in order to change the world.
The world needs changing,
but these large systems are emergent phenomenon,
they cannot be changed.

Meg Wheatley
Episode 090 Reclaiming Leadership for the Human Spirit

27

Agency = System Power

What powers a complex adaptive system? Agency basically means the power to affect a system. Agents are factors that cause a complex system to evolve through their interactions, or "co-evolve," since it is likely that the system also changes the agent. The more a "thing" can change a system, the more agency it has.

Agents in a system aren't always people. Sometimes they're concepts, or ideas, or subsystems. Sometimes the power to make ripples of change in a system is distributed among many factors—anything from networks of people, to the weather, to political messages. Agents can be relationships or shared stories. Even an idea can have agency. Ants, apparently, do not have agency—not the individual ants themselves—which makes them useful for studying complex systems.

To us, the fascinating thing about agency is that its power is derived from the way information flows through networks. So a person might say, "What does my opinion matter? I don't have agency in a system as big as national politics." The level of agency that individual human beings have is debatable. However, ideas spread faster through networks today than in any time in history. The power to change a system or an idea—or the world—can begin with individuals sharing information.

28

Data is Not the End of the Story

We systems thinkers, why do we so badly want to understand? Maybe we want to know the relationship of everything to everything else, of self to all-that-is. Maybe we want food on the table and world peace. For all of the answers, complexity is the network of networks—a very big idea that we now have better tools to explore. Yet data is not the end of the story. There is more.

We all have big questions and real-world problems that systems thinkers can only answer as a network, in love letters, if you will, between the heart and mind of all humanity. The perfect equilibrium is a moving target somewhere between data addiction and philosophical abstraction, where a better understanding of complex systems sparks new solutions to world problems. We suspect that complex systems weave through every living being and nonliving agent in every sort of ecosystem imaginable. Politics is one way things get done, but like data, it's just one facet of the big story of networks. Networks ensure our survival. They ensure there's room for everyone. We need philosophers, futurists, educators, and number crunchers in the conversation, exploring uncharted horizons and drawing new maps.

One of the insights I learned from Brian Goodwin is a complex dynamic system is any system with more than three interacting variables. It's governed by nonlinear mathematics and it's mathematically, fundamentally unpredictable and uncontrollable, unless you put a limited time boundary or space boundary around it, then we can do a bit of prediction and control if you throw lots of computing power at it. But, this deeper lesson that we are dealing with of uncertainty and uncontrollability makes you ask the question, how do you design for positive emergence?

Daniel Wahl
Episode 122 Design-Led System Transformation

30

Complexity Theory and Systemic Risk in Financial Markets

For Episode 123 Complexity Theory and Systemic Risk in Financial Markets, we talked with Francesco Filia, chief executive officer of Fasanara Capital, a boutique asset management company which focuses on niche investment strategies and researches systemic risk. Filia shared his perspective on the way financial markets are conventionally analyzed: "Analysis that is typically performed on financial markets, in my opinion, is quite outdated. It is still the same that we have seen in the last several decades. It has not evolved and could learn so much from nearby fields such as complexity theory." He described financial markets as ecosystems and continued to bring many different insights from complexity into the conversation, including how feedback loops in financial markets can create instability and lead to tipping points. Filia's passion for complexity science became evident when he shared, "Complexity theory has a number of teachings and tools that can help you hack your brain to see things more nonlinearly and to see things as a system, and not as the sum of its elements."

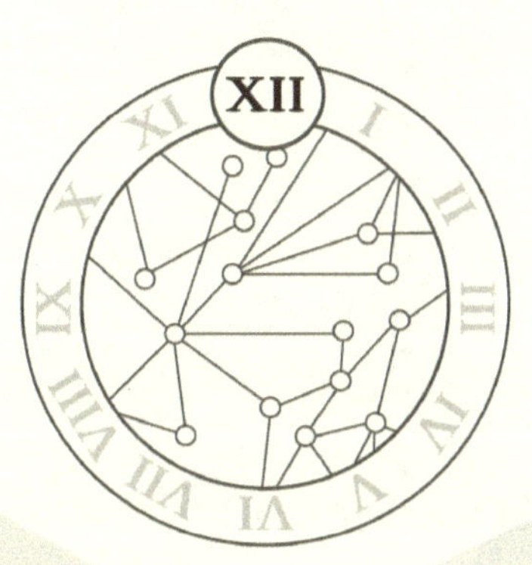

DECEMBER

1

Gratitude and Spirituality

"The angels were good to me."
Douglas Drane

Doug believes in angels. We knew that he talked to them often and listened to what they had to say. He was also incredibly grateful to them for his success, health, and happiness. With unfailing conviction, he talked about the "Big Angels"—also called archangels—referencing them by name and asking them for guidance. He often spoke of Uriel, the archangel of wisdom and light, who helps us connect to spiritual realms and discover our inner strength and higher levels of consciousness. When we asked Doug why he would reference the Big Angels rather than God, he told us that angels are the most universal, spiritual beings across the world and throughout many different cultures. His spiritual curiosity is both inspirational and contagious.

Doug is inexorably driven to ask big questions, like: What is my purpose? What is my charter? What is my role in changing the world? He pursued these questions relentlessly and taught himself how to listen deeply for the answers. He read countless religious texts, went to churches and meditation retreats, and explored the spiritual beliefs and practices of people around the world. One of his favorite books is *The Source Field Investigations*, by David Wilcock, and he would often reference the hidden energy fields that connect all of humanity as "the source field." We believe that Doug's love for complexity is an extension of his love for spiritual and scientific questioning. To Doug, complexity is more than a science, it is a way of thinking and being. For someone so spiritually and intellectually curious, only complexity could hold together everything that is important.

2

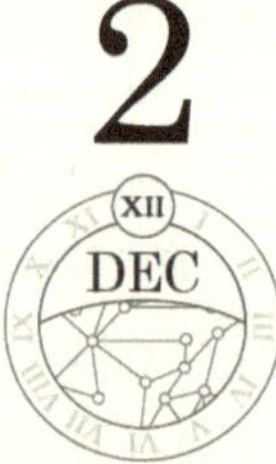

Feeling Complexity with Radical Wholeness

We talked with the developer of The Embodied Present Process, Philip Shepherd, for Episode 072 Feeling Complexity with Radical Wholeness. Shepherd said, "Every culture is a story about what it means to be human," and explained that Western culture has led us to believe in a story of separation and isolation. Shepherd urged us to learn how to feel and embrace our complex and deeply interconnected reality. "As you come back to the body and its intelligence, those rich complexities of love and fulfillment become available to you." Shepherd said, "There is more complexity in harmony than we could begin to analyze but, you can feel it, there are realms of complexity that reason cannot capture and pin down and analyze." This feeling sense of the embodied present is critical for personal growth and truly radical wholeness. As Shepherd said, "The body is our bridge to the present."

3

The Power of Perception

Network mapping can show the relationship of things to things, things to events, or ideas to actions. If nothing ever changed and you plotted your life on a timescale, it would be a flat line. Lack of diversity makes life without dimension, the opposite of dynamic.

If you looked at a line directly from its end, all you would see is a point, if you were lucky. Depending on your vantage point you can turn a straight line into a single point using perception alone.

When things do not change or our scale is wrong to detect them, we stop seeing them. Our eyes are trained to detect movement, all our senses honed to recognize the signals of change. Our very perception of time is relative to our experiences and is in part subconsciously determined. Do we come preloaded with a bias that tells us that if we have no means to measure a thing, it doesn't exist?

We don't know about you, but we've spent some time trying to keep things the same. It's an evolutionary survival technique, to be sure. Having to make the same decisions each day with different variables would eat up our mental resources, and we wouldn't have time to work our way up Maslow's hierarchy of needs. Change is constant, though. The rate of change may just be too fast or too slow for us to detect.

A system can move in the same way over and over for a long time—we call that periodic—and order is briefly rescued from entropy. We rely on those systems and delegate power to them. Other times we don't know when or how roughness will smooth itself into a pattern. It may approach zero and only really achieve it when the window of observation is small enough. We may need powerful telescopes, stop-motion photography, and long data modeling to make out the ends of what we are already sensing with intuition—whispered conversations with the fringes of our mental networks. The horizons of our perception may be so big that they appear as a straight line, but they are there, interrupting a shapeless, infinite sky.

4

It's the refusal of reductionists to look at the connections between things as being just as important as the separate parts themselves that cause, I think, so many problems. . . . We can think of complexity science and systems thinking as really the science of connection rather than the science of separation.

Jeremy Lent
Episode 074 The Systems Worldview of Human History

5

Cultivating Social Resilience

For Episode 021 Cultivating Social Resilience, we discussed how complex social problems influence our human networks and relationships with sociologist Josh Morgan. Morgan said, "I feel my purpose as a researcher is to answer questions about complex social problems and report on them as simply as I can." Morgan helped us explore different ways in which ordinary people can influence systems-level issues, like social trust. He said, "What makes humanity so interesting is that our complexities can't be explained with one theory because we're so highly adaptive." Morgan shared one of his favorite quotes with us, an ancient proverb which happens to also be one of our favorites: "A society grows great when its people plant trees whose shade they know they will never sit in."

6

Blindspotting

Often when we think about complex and global-scale problems, feelings of defeat can creep in on us. The scale and scope of the problems we face as a human race can feel very intimidating and it is easy to want to break each issue down into parts so they are more digestible and less overwhelming to tackle. The problem, though, is this reductionist approach does more harm than good because each problem is so intertwined with the next. We soon start to encounter unintended consequences from our actions because we are operating within a limited viewpoint and understanding of what is happening, and therefore we miss out on realizing what is necessary.

In Episode 054 Reversing Complex Problems for Global Security, James Greyson described a tactic he calls "blindspotting." Blindspotting is intentionally looking for what we've missed, pushing ourselves to imagine possibilities beyond the common debates. He uses blindspotting to "think bigger" and calls it the driving force behind all of his work. Greyson said, "Ultimately, blindspotting can address the neglected gap between the scale of our responses and the scale of the interconnected challenges."

7

One of the things that has dawned on me lately is that we really are just a constant state of change, which is crazy complex, and the second that we stop changing, is the day that we pass on.

Jason Dykstra
Episode 003 Using Systems Thinking to Resolve Conflict

8

A Holistic Perspective of Evolution

We spoke with theoretical biologist and author Josh Mitteldorf for Episode 110 A Holistic Perspective of Evolution. Mitteldorf discussed his work and shared, "I am working on how evolution holds in place traits that are good for the community in the long run, but they are bad for the individual in the short run." He discussed physics in the nineteenth century, saying, "It was thought to be mechanistic and reductionist . . . somehow you understand the little pieces and you put that together to make the whole, and physics had to abandon that with the advent of quantum mechanics." Mittledorf also talked about the evolutionary process, including gene sharing, cooperation, and natural selection. "Aging was my introduction into the world of group selection—what's good for the group, versus individual selection, what's good for the individual." All of this showed him, "Biology needs to embrace holism."

9

Magical Order

When we start geeking out with new complexity thinkers, we often hear the question, "What is a self-organizing system?" To respond, we like Diego Espinosa's description in Episode 006 Information, Ants, and Inequality, "A self-organizing system is basically a system in which order emerges from the different pieces following simple rules and interacting with each other."

It sounds kind of magical. The types of patterns that emerge from that self-organization are remarkable. It's all of evolution, it's ant colonies, it's human consciousness . . . that's what's so amazing about complex systems. Networks are complex in the richness of their interdependencies, and the pieces are all following simple rules.

10

Looking Back and Reflecting

In Episode 023 Throwback to 2015, we reflected on our *Human Current* journey of curiosity, learning, and exploration over the previous year. Co-host Angie Cross said, "We started this venture with curiosity. We thought, is this possible, to apply complexity and systems thinking to all areas of life and all fields? And we have learned a lot along the way." Co-host Haley Campbell-Gross added, "We have interviewed some great minds from fields of study like health care policy, education, entrepreneurship, marketing, finance, economics, and emergency planning."

During our casual conversation, we listened to a few clips and highlights from past guests, including Michele Battle-Fisher, Jason Dykstra, Isaac Morehouse, and Bonnie Caver. We compiled these clips to further explore the idea that a complexity thinking lens can be used in all aspects of life.

We shared a clip from Battle-Fisher defining a systems thinker. "First, a systems thinker needs to see the whole picture. . . . Number two, they're willing to change perspective. . . . Number three, we need to look at the interdependency of all the variables and if you are stuck into linear reductionism that's not very easy to come to terms with. . . . We have to pay attention to the long term. . . . Next, you need to think 'go wide' when you look at causes. . . . Focus on structure and not blame."

Similar to Battle-Fisher's definition of a systems thinker, Dykstra's said that systems thinking is "seeing the bigger picture and seeing how everything interconnects with one another . . . and part of that systems thinking is drawing the boundaries around it." When we asked Dykstra what someone can do to help shift company culture he responded, "One of the beautiful things about systems is that no matter where you are within that system, if you change, even just a little bit, that automatically forces everyone around you to change as well, because you're now interacting with them in a different way."

In a clip from Morehouse's interview, he discussed how we can get in the way of our own hopes and dreams. He explained that many people are "waiting for permission to do things—to create." He shared an example of people who want to be writers and asked, "What does it mean to want to be a writer?" Morehouse argued, "There are only two conditions you can be in. You can be a writer, which means someone who writes regularly, or you can not be a writer, which is someone who doesn't." For people who say they want to be a writer, he would ask, "What are you waiting for?" People have responded to Morehouse with, "Well, I want to get better first," to which he responds, "How do you get better at writing?"

Finally, we shared a clip of Caver's response to our question, "How do you use systems thinking in reputation management? How does thinking about brands and reputations in terms of systems and ecosystems help you in your work and help a company build a healthy brand?" Caver responded, "It's really important to look at what is in an ecosystem of a reputation and a brand and a culture. And so, we really look at culture, brand, and reputation as being interconnected. If you have a good culture it allows you to deliver on your brand promise and build a brand experience and equity with your stakeholders, and then you can execute toward a brand vision. . . . In order to do that, if you are looking at your ecosystem of your company the first thing to do is break down silos. You can't really look at the whole vision of a company by having silos."

At the time we didn't know *Human Current* would reach well over 100 episodes. But after only twenty-three episodes, we were confident that we could answer our question in the affirmative: yes, it is possible—and a lot of fun!—to apply complexity and systems thinking to all areas of life and all fields of study.

11

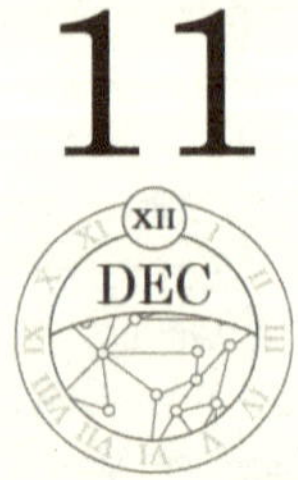

I think from the application point of view, what we are trying to learn is how this collective motion or group motion [flocking] behaves as an organism. Sometimes groups, like millions of birds, don't crash, so to understand what kind of interactions govern the development of this collective motion is important.

Victor Dossetti
Episode 111 Stories and Insights from the Complexity Community

12

Turning Affordable Housing on Its Head

We talked with Walter Moreau, the executive director for Foundation Communities, a nonprofit that creates housing where families can succeed. In Episode 022 Turning Affordable Housing on its Head, Moreau shared that an important function of his career of twenty-five years has been to educate people who have prejudices or mental models toward residents of affordable housing.

"The immediate image many people get in their mind is the projects, and that means that it's crime, and drugs, and lower property values, and ugly," he said. Moreau wants to change that. "Affordable housing can and should be a foundation for families, an opportunity for kids to get a good education and put health services right at people's doorsteps. We're turning on its head this whole notion of what affordable housing can be."

Moreau described a whole-systems or whole-person approach, and why it is important. This approach goes beyond simply providing housing. A whole-person approach includes health and wellness, education, feeling safe, and being part of a community. Moreau described how Foundation Communities leverages a systems approach to affordable housing in order to provide a successful, sustainable program for families. He said, "Beyond systems thinking, [approaching affordable housing] is more chaos theory. You have to be able to roll with the ups and downs of development." We were inspired by Moreau's stories and perspective, and especially his firm belief that "affordable housing isn't for 'those people,' it's for us—people we know."

We learned that one of Moreau's favorite books is also one of ours, *Harnessing Complexity* by Robert Axelrod and Michael Cohen. Moreau shared his favorite quote from this book, "It is often tempting to assume that the agents of a system are all basically the same."

13

When you ask a question and you get an answer,
really be critical of, is that answer enough?
Is that the answer that everyone's always said,
or could there be more lurking variables which impact that?
And when you start to ask those questions
you start to see those variables.
The system becomes more complex
and therein lies opportunity for creative research
and creative business opportunities.

Brad Glisson
Episode 113 Connecting with Complexity Thinkers

14

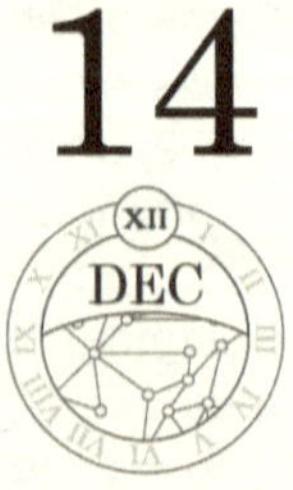

Blockchain

However complex our systems are, there will always be people trying to exploit weaknesses in systems—and people attempting to make them more secure. Blockchain technology is an example of an extremely sophisticated method of providing security in digital transactions, linking lists of records (blocks) using cryptography.

For Episode 111 Stories and Insights from the Complexity Community, we asked Roozbeh Daneshvar, blockchain software engineer at IBM, about his thoughts on blockchain and how it will be used in the future. Here is what he had to say: "There is a perception that it [blockchain] is going to fix everything, and we will all be happy ever after, but this definitely depends on the use case and how we are going to use it, and to be honest I don't know. I don't know where we are going. What is going to happen still depends on applications, use cases, how we take advantage of this opportunity. The idea is simple but it's what the applications are going to bring that make it more interesting. Cryptocurrencies, that's one of them and it seemed that maybe exchanging cryptocurrency without the blockchain would have been maybe impossible, because cryptocurrency is not something physical, it's either you know about it or you don't. This is one use case but there might be other use cases; for example, let's say we want to vote together, we might leverage something like a blockchain to record these votes and who voted for whom."

15

Learning from Social Networks

In Episode 111 Stories and Insights from the Complexity Community, we talked with Rosa Benito, physics professor at the Universidad Politécnica de Madrid, who shared how we can learn from social networks. Benito described the opportunity saying, "We hope to have a view of human behavior through the use of social media. . . . We have a huge opportunity to analyze data that comes from the behavior of people and it can tell us a lot about the way to communicate their ideas." She described the example of political campaigns, explaining that there is vastly more information available about people's points of view in social media than there is from polling and traditional opinion research methodologies.

16

Understanding Complexity in Context

Episode 031 Understanding Complexity in Context was a reflection episode where we shared clips from past guests: Gloria Burgess, T.K. Coleman, and Jason Dykstra. We often asked our guests how they became interested in systems thinking or complexity theory, and we were intrigued by their responses. Many of our guests shared a common theme of realizing they were a systems or complexity thinker since childhood.

For example, Burgess described how she grew up in a complex system and explained, "I was born in the rural south in Mississippi and I was born into a very segregated community. I didn't know it was segregated. I thought I grew up in an all-black town, honestly, and that once a week my family would go into this other place where mostly white people were and I thought that they were two separate towns, but they were actually the same town. . . . When I talk about growing up in this complex system, my mom had seventeen brothers and sisters so I had aunties and uncles, lots of cousins. . . . I had a village just within my family, but like I said, also the community members took a vested interest in raising me and my sisters . . . so when I think about that kind of environment, I think about my family as my first complex system and understanding the rules, like, who had power? Who had rank? Who had status? Who had privilege? I work with these kinds of things in my professional life, in terms of unpacking systems so that we can figure out how things work and how to make things different or better or more equitable."

Another question we often asked our guests was why they believed understanding complexity mattered. In Coleman's clip he responded, "Understanding context and what systems theory has to say can shine a light on apparent discrepancies." When we asked Dykstra the same question, he said, "Understanding how we can approach complexity is so useful in our everyday life. Complexity is all around us. It's a great opportunity to learn more about ourselves and our role within the world."

17

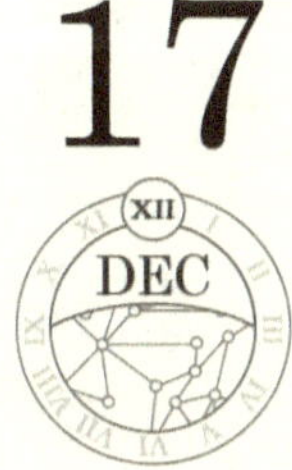

Worldwide Networks

At the International Conference on Complex Systems 2018, we talked with Javier Borondo, a research scientist and chief technology officer of AGrowingData. In Episode 111 Stories and Insights from the Complexity Community we shared his insights about the food and agriculture industry as a complex system. "If you think about it, the food industry and the farming industry are interconnected over the world. It's a complex system, where what is happening today, for example, to the farmers of Spain is affected by what happens in Africa, or other parts of Europe, or here in the USA. [Another example], Italy's economy may not either import or export too much, but if you get the wrong climate over there and they are not able to grow vegetables for themselves, then they will go out to the world market and they will just make a big break in the market. If you study Italy by itself or Spain by itself, you would never be able to capture this, but whenever you study the whole agriculture sector like a complex system, then you get an understanding of the way it is happening. This is really important, because it's a threat not to be able to produce enough food for the whole population in the next years."

18

Connecting with Complexity Thinkers

We shared four of the interviews we recorded at the International Conference on Complex Systems 2018 for Episode 113 Connecting with Complexity Thinkers. These interviews were with Brad Glisson, Liz Johnson, Vinesh Raja, and Rhys Lindmark. Each of their personal stories wove together some of their shared values, ideas, and visions for the growing field of complex systems science.

Glisson, co-founder of Thoughtblox, discussed his work and interest in information. "I'm really passionate about how information is able to grow since, as humans, we can codify, code, and create artifacts of intelligence."

We asked Johnson, managing editor for the *Journal on Policy and Complex Systems*, what advice she would give people who are looking to explore complexity science. She said, "It takes a natural curiosity. There are now college courses. Universities are offering classes on networks and complexity science—some that aren't as mathematically oriented, which makes it more amenable for different types of people."

Raja, bioinformatician and software engineer, shared his enthusiasm for *Human Current* and complexity thinking. "I've been listening to a lot of *Human Current* podcast [episodes], and that's how I got to know about the conference. It's been exciting studying about complexity . . . it's just now that complexity is coming up a lot these days."

Lindmark, a fellow podcaster whom we met at the conference, is the head of community and long-term society impact at Massachusetts Institute of Technology Media Lab. "The work that I do is based around this macro-societal transition that we're going through," Lindmark said as he discussed his work. "We're at the end of the Industrial Revolution, late-stage capitalism, the start of this information revolution in the digital age, and it's going to be a fragile time because we have four billion network smartphones, and that allows you to learn, create, and distribute value a lot more quickly. So that's why we have massive, converging, exponential technological change."

19

Innovation has no age, only productivity does.
That is the reason you see so few people
making breakthroughs later in their career,
but they do exist.
It's not because people later in their career are not creative,
but they are not productive. They stop trying.
Those who continue to try
will be just as successful as the young individuals.

Albert-László Barabási
Episode 115 The Network Science of Success

20

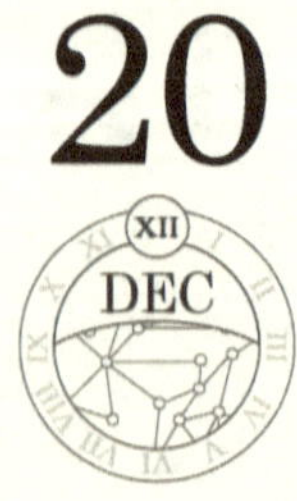

Why the Universe Hates Stagnation

The universe abhors too much sameness. When you sense that parts of you or your efforts are stagnant, or not synthesizing new experiences into your character or your work, do you notice those subtle feelings of self-loathing? It happens the same way in any network: a culture, a household, a team, or a country. Individuals turn against their own identity structures. Cancers form. Relationships become a battle of projections. Societies polarize. The eventual result is change.

Proximity creates pressure. When the pathways of a network are limited, pressure builds up and diversity, essential for a complex system's survival, disappears. Movement is limited. Tension increases. When these things happen, it seems as if there are limited channels through which energy and information can flow, and limited combinations of ideas that can be drawn upon to solve problems.

The nature of stagnation is too much order, failure to adapt, an endless replication of "sameness" that perpetuates the same unsolvable paradox of existence without new understanding. The question is, if nature sees destruction as an effective means of creation on *all* scales, at what scale will we human beings consciously allow adaptive and even destructive processes to evolve us? How can we architect these natural processes into our projects and lives?

21

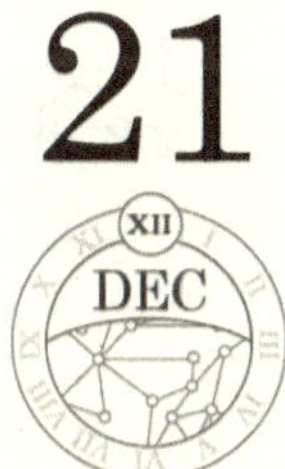

The Usefulness of Reflection

In Episode 071 Throwback to 2017, we reflected on the ground we covered in 2017, including visionary leadership, learning from nature, environmental consciousness, global security, brain science, human networks and interconnectedness, happiness, and meaningful work.

Co-host Haley Campbell-Gross said, "I think it's really important to have these moments to pause and reflect, and there's no better time than at the end of the year." She shared a quote by Meg Wheatley, "Without reflection we go blindly on our way creating more unintended consequences and failing to achieve anything useful." We also announced exciting news about upcoming guests for 2018, which included Melanie Mitchell from the Santa Fe Institute and Jean Boulton, co-author of *Embracing Complexity*. We were ready for anything in 2018!

22

Intention

We set out an intention from the very beginning—we talked about it during Episode 001 Why Complexity?—that our conversations on the *Human Current* podcast would explore how learning complexity could help us work and live happier. We didn't have the perfect language, but we wanted to build our vocabulary for complexity awareness and tolerance in all walks of life. That's where our guests came in—our curiosity, coupled with their expert knowledge, made our show a rich journey we could all learn from.

23

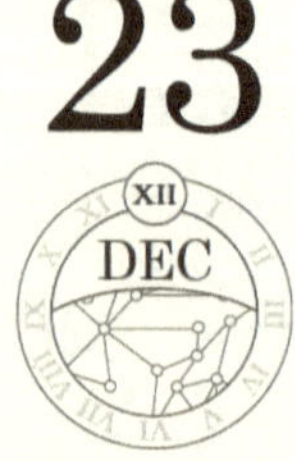

The Spaghetti Marshmallow Challenge

Let's talk about the f-word.

We dedicated Episode 002 The F-Word to a lighthearted conversation about one of our greatest fears: failure. We've noticed that we don't start out afraid of failure as kids. And we can prove it—with spaghetti and marshmallows. Stanford University's d.school, among other design schools, has run the Spaghetti Marshmallow Challenge with teams of five-year-old kids and adult work teams from organizations. They give the teams a bunch of uncooked spaghetti noodles and some marshmallows and ask them to build a stable structure within a constrained period of time.

The adult teams tend to spend time planning, strategizing, assigning roles . . . all behaviors intended to avoid failure. The kids, on the other hand, just dive in. And the kids win, nearly every time. They create many more "failed" iterations than the adults do in the time allowed, and end up with a better result. They also laugh more and have more fun.

Successful teams come from a place of curiosity and play. If their noodles break, they just begin again. If they eat a marshmallow, they decide they didn't need it anyway.

The Spaghetti Marshmallow Challenge teaches us that sometimes you just have to get your hands sticky, eat some marshmallows, and break some spaghetti noodles. Failure can be fun.

24

The Mystical Nature of Wholeness

We continued to learn from others about interdependencies, patterns, relationships, emergence, self-organization, feedback loops, and many more hallmarks of complex adaptive systems. All of these tangible concepts helped us better understand the inner workings of our reality. However, there was another concept we found more elusive to grasp with the mind: wholeness. Wholeness is the very essence of what it means to be a living system.

At first thought, it seemed like a straightforward term that we could wrap our heads around without having to sit with it for very long. However, when we did sit with it for a while, we realized it is loaded with mystical complexity.

We started reading everything we could about wholeness as well as talking about it with our guests. Here are just a few of our favorite insights about wholeness:

> "The double role of living systems as parts and wholes requires the interplay of two opposite tendencies: an integrative tendency to function as part of a larger whole, and a self-assertive, or self-organizing tendency to preserve individual autonomy."
> Fritjof Capra
> *The Systems View of Life: A Unifying Vision*

> "We are not broken people. It's our relationships that need repair. It's relationships that bring us back to health, wholeness, holiness. . . . When we seek for connection, we restore the world to wholeness. Our seemingly separate lives become meaningful as we discover how truly necessary we are to each other."
> Meg Wheatley
> Episode 090 Reclaiming Leadership for the Human Spirit

"Individuality is only possible if it unfolds from wholeness."
David Bohm
In Renée Weber, *Dialogues with Scientists and Sages: The Search for Unity*

"Out of the cacophony of random suffering and chaos that can mark human life, the life artist sees or creates a symphony of meaning and order. A life of wholeness does not depend on what we experience. Wholeness depends on how we experience our lives."
Desmond Tutu
Desmond Tutu Peace Foundation

In *The Fifth Discipline*, Peter Senge wrote, "Systems thinking is a discipline for seeing wholes. It is a framework for seeing interrelationships rather than things, for seeing patterns of change rather than static snapshots." Wholeness is all around us. It is our reality. We can use systems thinking and complexity thinking to see it more fully and tap into it. We need this discipline in order to fully appreciate what wholeness really means.

25

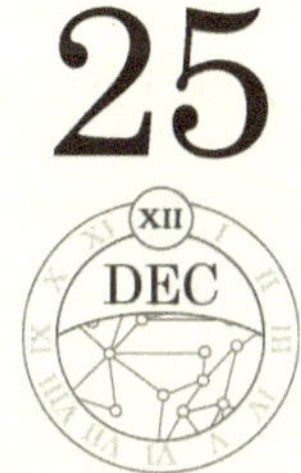

Human Networks Inspire Holiday Cheer

Episode 045 Human Networks Inspire Holiday Cheer was our last episode of 2016, so we took some time to reminisce about the past and explore the power of interconnection and networks. In our conversation, we discussed how the holidays can be a difficult time for many people. "Some people really struggle with finding that connection and support that they need. And so, it's more important than ever to discuss human networks and the importance of connection," said co-host Haley Campbell-Gross.

We shared clips of three guests from our series on human networks—Mary Ann Herrick, Jasper Faolan, and D'Artagnan Caliman—who responded to the following quote from the book, *Can't Not Do: The Compelling Social Drive That Changes Our World*, by Paul Shoemaker: "I do not think our biggest social challenge is know-how, it's not even money. It's connecting the right people, with the right ideas, in the right way."

Herrick responded, "That's really powerful because it is easiest for us to jump to the conclusion that if people had the right knowledge, they would make the right choices, or if only we had enough money we could do what was needed, but perhaps those things already exist the way they need to be, and if we were just able to connect people together who have those things, then there could be an incredible impact." Herrick and co-host Angie Cross discussed the power of creating a groundswell. With social movements, "The strength is in people, and the strength is in numbers, and it's definitely the human network and maybe even the human current." Herrick responded with, "Power in coming together, that's for sure!"

Faolan said, "This is a beautiful quote," because it speaks to how the inanimate objects of our system are not what makes impact. "Those things are not what changes the world, those things are only tools." Faolan stated, "Paul Shoemaker is talking about how to be able to drive social change. And the way that we really do that is not in individual silos, but to be able to reach out, using our personal narratives to go out into the world, to speak our truth and say what it is that we want to see changed."

Caliman said that far too often we think about the financial resources that we need to have "to really serve families or to improve the lives of people." Caliman went on to say, "It really just takes someone caring and listening and reaching out to an individual or family to help them." While it's important to think about the necessary resources for helping communities, families, and individuals, Caliman also suggested that we think about the personal connections and "what we can do as humans to be giving of ourselves." He concluded with, "That's what that quote means to me. I think it is so important and I think that sometimes we run the risk of losing sight of that, so it really is about the human connections. It really is for us to stand up and be that safety net for individuals, families, and community."

26

There Is No Certainty

You can dig quite a rabbit hole with the phrase "everything is connected to everything else." However, it is safe to say that nothing in nature exists separately from its environment or its neighboring phenomena, so in some way, everything is in motion. On top of that, the way that we move has actual bearing on how we see and measure the world and all its systems of movement. That's some special relativity!

Seeing in networks makes words like "prediction" and "certainty" catch on our tongues. From a networked frame of mind, laws and absolutes become contextual when considered on a certain scale, as they too are networked and subject to emergence.

We're saying some preliminary goodbyes to certainty. In a way, that makes us feel more secure. How heavy a burden would it be to have solved the mysteries of the universe using laws set in stone, and still not have solved its "problems"? Even stones are broken and smoothed by time. Which phrase makes you feel better: "Nothing is permanent," or, "Everything can change"? We think of Lewis Carroll's Alice from *Alice in Wonderland*: "It's no use going back to yesterday, because I was a different person then."

27

XII DEC

@LetsWorkHappy

For Episode 019 Let's Work Happy! we explored what happiness at work meant to the members of the *Human Current* team. Continuous learning and personal growth were values we practiced at *Human Current*. One of the ways we lived out these values was through having a book club at our office. We would commit to reading a book, often one authored by an upcoming *Human Current* guest, and once a week we would discuss the chapters assigned for that week. One of our earliest book club books was by Shawn Achor, *The Happiness Advantage.* We thought this book was fitting for our book club, especially with our Twitter handle of @LetsWorkHappy, and because we were often asked what happiness has to do with complexity, networks, and ecosystems.

One of the things we learned from Shawn Achor's book was, "Seventy-five percent of job successes are predicted by your optimism levels, your social support, and your ability to see stress as a challenge instead of as a threat."

Learn more about some of the books we explored for our book club from the Bookshelf at the back of this book.

28

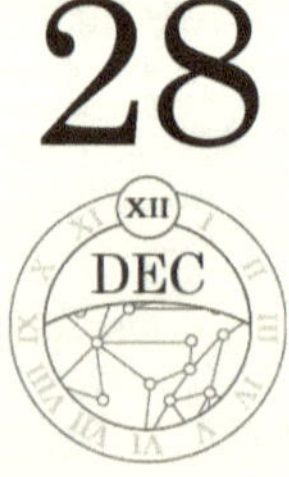

A Conversation About Wine

You can imagine our delight when we chatted with Albert-László Barabási and the conversation turned to wine for Episode 115 The Network Science of Success. Here is an excerpt from that conversation:

***Human Current*:** I especially loved your description of the first law [in your book, *The Formula*], when you talked about wine. Good wine is just good wine. [Laughter.] It's hard to tell which is better and for judging, there's not really a scale there to measure good wine.

Barabási: Absolutely. Wine has always been a big frustration for me because I am a scientist and I like universal things. I always felt that wine is not universal enough. There's always a different year that tastes differently and always a different maker, so that we can never make sense of it. . . . There's this great piece of research to show that it is not only me who cannot really capture the good wine. Not even the top experts, at the top competitions in California, are really objectively able to tell apart good wine from good wine. Once again, even a non-expert can easily tell the difference between bad and good wine. Once you put three good wines in front of me, I'm unable to tell the difference. It turns out to be not only me, nor can the experts.

***Human Current*:** I want to know who these people are that get paid to judge wines? [Laughter.]

Barabási: Well, it's like everything else. It sounds like lots of fun that your job is to judge wines. I actually think that my job is even more fun than that, I'm getting paid to actually do research. I think what the key is for success is to find a happy moment in whatever you do and not rely on external recognition to justify your choices.

29

Reflection

by co-host Haley Campbell-Gross

I remember the very first day I met Doug, and thinking he was either totally crazy or totally brilliant—only to learn shortly after, he is both! He is brilliant because he is so crazy and vice versa. He thinks of the world and humanity as interconnecting fields of energy and sees the potential in everyone and everything. His support and mentorship during my time with HPT Development, home of *Human Current*, was a precious gift that I knew would be special from the day we met.

I was with my HPT family for almost six years and during that time I got married, had my son Jack, and bought a house in the suburbs of Austin, Texas. I grew immensely both personally and professionally because I had the privilege to do work that I loved with people that I loved. Doug and my co-workers truly were my second family. They encouraged me to work on projects that brought me joy and challenged me. A few years into my time with HPT, an opportunity came along to help the *Human Current* team.

When I began working with *Human Current* shortly after the launch of the podcast, I knew the project would change my life. I remember feeling deeply inspired every day by a guest we talked with, something we read, or a conversation with my co-host Angie. Angie and I became very close over the years and shared many wonderful evenings sipping wine and laughing (or crying) with each other. We had such a special working relationship and will continue to be close friends. Our love for each other and our work made the podcast a welcoming space for people to share their time and wisdom. I will forever be grateful for the time I got to spend working with our *Human Current.*

30

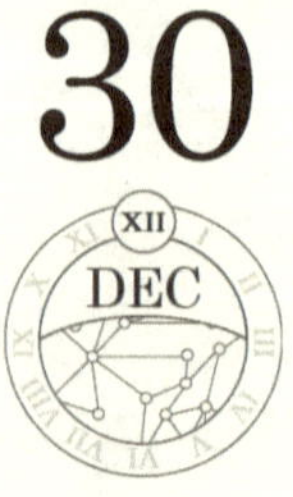

Memories

by co-host Angie Cross

I worked with Doug for more than six years, and for most of that time we were working on *Human Current*. While my work with Doug and *Human Current* has wrapped up, the relationships, learning, and memories will always remain with me, and as a result I am forever changed and grateful. We often referred to our work with Doug as a unicorn job because it was magical, rare, and like a fairytale. When we talked about how unique and special our experience was, people might not believe us. But every little bit of our unicorn was true.

Those six years were filled with pivotal times in my life—big life milestones. My son, then nineteen years old, moved out on his own; I earned my bachelor of arts in organizational communication; I earned my master of arts in leadership; my husband lost his father and I lost dear friends, including my dear sister-friend Misty, which still leaves me heartbroken; and most recently, my husband and I sold our home, bought some land, and built our sweet ranch home. There were highs and there were lows, and through it all, because of where I worked and the people I worked with, I could learn and grow. Our work environment was a place of creativity, learning, vulnerability, and authenticity. When we showed up for work, we were expected to bring our whole selves. We celebrated each other, we leaned in to learn from each other, and we were understanding of each other.

There were those special traditions and rituals that we created, like having weekly book club. We had a salad club where we all brought different ingredients each week to make and eat salads together every day (yes, every day!). Participating in holiday food and toy drives, volunteering for Meals on Wheels and Habitat for Humanity, celebrating birthdays and other special moments . . . these traditions and rituals helped foster our unicorn-like work environment and culture.

I will cherish the relationships, many of which I know will be lifelong, and so many wonderful memories. I have memories of learning from each other and from reading countless books, so many deep conversations, and belly laughter . . . lots of belly laughter. It's no wonder *Human Current's* Twitter handle was @LetsWorkHappy. Often when we recorded our casual conversations we had to do several takes and I made tons of edits because, without fail, we'd start laughing about something (often at ourselves) and couldn't keep it together without giggling. Editing the takes would kick-start my laughter all over again. I am sure that sipping on wine while we were behind the mics probably didn't help us be serious and on-point some evenings, either!

As I reflect on this time in my life, I feel so grateful. It was a gift to be able to start the podcast with Lindsay Paterson and Stacy Hale, and then having Haley Campbell-Gross join the podcast was the perfect fit. Working with Haley never felt like work. I'm sure Doug would say that's because we're so aligned. The friendship that Haley and I have developed is much like the unicorn place we worked—it's rare, special, and one of a kind. I will treasure this relationship forever.

31

Gratitude

It has been such a gift to have the opportunity to interview so many brilliant people who have such big hearts and are crazy-smart. We could feel the passion these thinkers have for complexity, systems thinking, wholeness, and our shared humanity.

We took to heart the guests and listeners who gave us feedback, ideas, suggestions—and those who expressed their appreciation. There were so many guests who shared their gratitude with us about the work we were doing to share stories about complex systems.

One moment of gratitude we remember was from our guest Michele Battle-Fisher who said, "We are not from the outside of the system. I think it's really unfortunate that being in complexity and systems we often have to seek out license or permission to voice this perspective. And so, this is why the [*Human Current*] podcast is so important. I hate to use the term normalizing, but we need this podcast and getting more individuals to see that this is a legitimate way of viewing the world."

Tanuja Prasad said, "First of all, I'd like to thank you at *Human Current* for what you do. Your podcast is both serious and casual, and it's a particularly wonderful way to present science. . . . It's an interesting time to be alive. There are so many dramatic changes going on and we're all experiencing them, but at the same time there's this power we have at our hands today."

Another heartfelt moment was when Gloria Burgess said, "This topic, the complexity of systems, institutional racism, all of these things are so important, and they're not the main thing—they're not the main event for most people. . . . So what you're doing, Angie and Haley, is just phenomenal, to be able to lift the conversation up and out and get more people involved, get more lenses to look at these issues."

These moments and many more of gratitude from our community of guests and listeners made our journey incredibly enriching and fulfilling.

Cheers to complexity, and cheers to our human current. To Doug, to our guests, to all our listeners and supporters, and to *you*: Cheers!

Human Current *co-hosts, Haley Campbell-Gross (left) and Angie Cross (right), with Douglas Drane.*

Wine Rack

We have many things in common with our mentor Doug, including our love for wine. We love that having a glass of wine with Doug starts with raising our glasses and giving cheers to something or someone. For those who have listened to our casual conversation episodes, you have heard us mention the wine we were drinking while we sat across the mics from each other and discussed the topic of the day. The following list captures the wines we sipped on while we recorded our casual conversations:

Ancre Hill Estates Pinot Noir, Wales, 2013

B.C. Bubbly Sparkling Wine, Okanagan Valley, Canada, 2011

Cantine due Palme Pillastro Primitivo, Puglia, Italy, 2016

Château La Loubière, Bordeaux Supérieur, France, 2012

Château Le Coin, Sauvignon Gris, Bordeaux, France, 2011

Cooper Point Merlot, Paso Robles, 2013

El Bombero Gran Reserva, Cariñena, Spain, 2009

Giesta Dão, Portugal, 2015

Grande Fleur de Lyeth Reserve, France, 2016

Grignolino d'Asti, Italy, 2012

Hacienda De Luna Merlot, Valencia, Spain 2013

Il Primo Selezionata di Alberto Romeo, Sangiovese-Malbec, Argentina, 2015

Integral Cabernet Sauvignon-Merlot-Cabernet Franc, Bulgaria, 2014

J Opi a.k.a. Rodolfo Sadler Malbec, Argentina, 2018

Paul Mas Estate Pinot Noir, France, 2015

Principe Strozzi Selvascura, Tuscany, 2014

Quails Gate Merlot, Okanagan Valley, Canada, 2013

Rotari Brut Tronto, Italy, 2013

Saracosa Toscana Rosso, Italy, 2016

Tenuta di Somaro Puglia Aglianico-Primitivo, Italy, 2014

Torrevento Primo Salice Salentino Rosso Riserva, Italy, 2008

Villa Puccini Toscana, Italy, 2012

Waxed Bat Reserve Cabernet Sauvignon-Malbec, Argentina, 2014

Willy Willy Shiraz, New South Wales, Australia, 2014

Cheers to our *Human Current*, and cheers to Doug for his mentorship!

Bookshelf

Doug loves books and shares this love with those around him. The walls throughout the office are lined with books, and there are multiple copies on hand of many of the books so people can help themselves. Doug even had a sign made for the bookshelves that reads:

Please, feel free to take a copy of a book!
Learn more, be more, change the world!

Doug has a book for every occasion, and he'd often assign specific books for us to read. We also love books, so we started a weekly book club in the office where we would all read a book and discuss it over lunch. This was a highlight for us and, of course, for Doug.

In the spirit of Doug's love for books and his expansive book collection, we thought it was only fitting to include a "bookshelf" section in this book. Along with the books and resources that our guests have written and recommended, the following are other go-to books on our bookshelves, and on Doug's.

Achor, Shawn. *Big Potential: How Transforming the Pursuit of Success Raises Our Achievement, Happiness, and Well-Being*. 2018.

Baron, Renee, and Elizabeth Wagele. *The Enneagram Made Easy: Discover the 9 Types of People*. 1994.

Brown, Brené. *Dare to Lead: Brave Work, Tough Conversations, Whole Hearts*. 2018.

Fields, Jonathan. *Uncertainty: Turning Fear and Doubt into Fuel for Brilliance*. 2011.

Freedman, Lawrence. *Strategy: A History*. 2015.

Gelb, Michael J. *How to Think Like Leonardo Da Vinci*. 1998.

Goldschneider, Gary, and Joost Elffers. *The Secret Language of Relationships: Your Complete Personology Guide to Any Relationship with Anyone*. 2013.

Goleman, Daniel, Richard E. Boyatzis, and Annie McKee. *Primal Leadership: Unleashing the Power of Emotional Intelligence*. 2013.

Hawkins, David R. *Power vs. Force: The Hidden Determinants of Human Behavior*. 2014.

Hofstadter, Douglas R. *Gödel, Escher, Bach: An Eternal Golden Braid*. 1999.

Hoverstadt, Patrick, and Lucy Loh. *Patterns of Strategy*. 2017.

Jaques, Elliot. *Requisite Organization*. 1996.

Kabat-Zinn, Jon. *Wherever You Go, There You Are: Mindfulness Meditation for Everyday Life*. 2005.

Meadows, Donella H, and Diana Wright. *Thinking in Systems: A Primer*. 2008.

Pentland, Alex. *Social Physics: How Good Ideas Spread—The Lessons from a New Science*. 2014.

Pert, Candace B. *Molecules of Emotion: Why You Feel the Way You Feel*. 1999.

Rath, Tom. *Strengthsfinder 2.0*. 2007.

Sachs, Jonah. *Winning the Story Wars: Why Those Who Tell—and Live—the Best Stories Will Rule the Future*. 2012.

Stroh, David P. *Systems Thinking for Social Change: A Practical Guide to Solving Complex Problems, Avoiding Unintended Consequences, and Achieving Lasting Results*. 2015.

Thurner, Stefan, Rudolf Hanel, and Peter Klimek. *Introduction to the Theory of Complex Systems*. 2018.

Wilcock, David. *Source Field Investigations: The Hidden Science and Lost Civilizations Behind the 2012 Prophecies*. 2011.

Contributors and Episode Resources

Episode 000 Welcome to Human Current
Angie Cross (see "About the Authors")
Stacy Hale, co-founder of *Human Current*, journalist, and complexity nerd

Episode 001 Why Complexity?
Angie Cross (see "About the Authors")
Stacy Hale (see Episode 000)
Bolman, Lee G., and Terrence E. Deal. *Reframing Organizations: Artistry, Choice, and Leadership*. 2017.

Episode 002 The F-Word
Angie Cross (see "About the Authors")
Stacy Hale (see Episode 000)

Episode 003 Using Systems Thinking to Resolve Conflict
Jason Dykstra, conflict resolution specialist and international speaker
jasondyk.com
Stone, Douglas., Bruce Patton, and Sheila Heen. *Difficult Conversations: How to Discuss What Matters Most*. 1999.
Frankl, Viktor Emil. *Man's Search for Meaning*. 2006.
Dweck, Carol S. *Mindset: The New Psychology of Success*. 2008.
Dweck, Carol S. "The Power of Believing That You Can Improve." *TED Talks*. 2014.

Episode 004 Complexity and Flow
Angie Cross (see "About the Authors")
Stacy Hale (see Episode 000)
Csikszentmihalyi, Mihaly. *Flow: The Psychology of Optimal Experience*. 2008.

Episode 005 Culture and Human Systems at Work
Angie Cross (see "About the Authors")
Stacy Hale (see Episode 000)

Episode 006 Information, Ants, and Inequality
Diego Espinosa, former money manager and founder of complexity-based Sistema Research
evonomics.com/author/diego
Espinosa, Diego. "How Ants Solve the Inequality Problem." *Evonomics.com*. March 7, 2016.

Episode 007 Certainty Merchants
Diego Espinosa (see Episode 006)

Episode 008 Trust the Process
Angie Cross (see "About the Authors")
Stacy Hale (see Episode 000)

Episode 009 Systems Citizens
Kacem Habiballah, program head of the Avionics Department at British Columbia Institute of Technology
bcit.ca/contacts/kacem_habiballah
Cathy Sturgeon, co-founder and principal of Work Innovation Partners
Both guests earned their Master of Arts degree in Leadership from Royal Roads University. royalroads.ca

Episode 010 Super Education
Douglas Drane, founder of Atex and co-founder of *Human Current*
Senge, Peter M. *The Fifth Discipline: The Art and Practice of the Learning Organization*. 2006.

Episode 011 How to Be Your Own Resume
Isaac Morehouse, writer and founder of Praxis
isaacmorehouse.com
discoverpraxis.com

Episode 012 Systems Thinkers Share Their Stories
Alicia Barron, co-founder of Caboh Collaboration Software
Khotso Khabele, co-founder of KoSchool
Michele Battle-Fisher, public health policy scholar
mbattlefisher.com
Battle-Fisher, Michele. *Application of Systems Thinking to Health Policy and Public Health Ethics: Public Health and Private Illness*. 2014.

Episode 013 Public Health as a Complex System
Michele Battle-Fisher (see Episode 012)

Episode 014 Wellness is Contagious
Angie Cross (see "About the Authors")
Stacy Hale (see Episode 000)

Episode 015 Why Complexity? A Thursday Throwback Reflection
Angie Cross (see "About the Authors")
Stacy Hale (see Episode 000)

Episode 016 The New Reputation Economy
Bonnie Caver, president of Reputation Lighthouse
replighthouse.com

Episode 017 Emergent Ecosystems
Angie Cross (see "About the Authors")
Stacy Hale (see Episode 000)
Hurst, David K., and Brenda J. Zimmerman. "From Life Cycle to Ecocycle: A New Perspective on the Growth, Maturity, Destruction and Renewal of Complex Systems." *Journal of Management Inquiry*. December 1994.

Episode 018 Preparing for Complexity in Emergency Management
Thomas Appleyard, manager of planning and programs with Ontario's Ministry of Health and Long-Term Care

Episode 019 Let's Work Happy!
Angie Cross (see "About the Authors")
Stacy Hale (see Episode 000)
earth.columbia.edu
worldhappiness.report
Achor, Shawn. *The Happiness Advantage: The Seven Principles of Positive Psychology That Fuel Success and Performance at Work*. 2010.
McKee, Annie. "Being Happy at Work Matters." *Harvard Business Review*. November 14, 2014.

Episode 020 The Energy That Binds Us
Josh Morgan, sociologist and podcaster
joshcmorgan.com

Episode 021 Cultivating Social Resilience
Josh Morgan (see Episode 020)
Morgan, Josh. "The Decline of Trust in the United States." *Medium*. May 20, 2014.

Episode 022 Turning Affordable Housing on Its Head
Walter Moreau, executive director for Foundation Communities
foundcom.org
Axelrod, Robert M., and Michael D. Cohen. *Harnessing Complexity: Organizational Implications of a Scientific Frontier*. 1999.

Episode 023 Throwback to 2015
Michele Battle-Fisher (see Episode 012)
Jason Dykstra (see Episode 003)
Isaac Morehouse (see Episode 011)
Bonnie Caver (see Episode 016)

Episode 024 Discover Your Intercultural Intelligence
Gloria Burgess, motivational speaker and leadership coach
gloriaburgess.com
The Human Library. humanlibrary.org
Detweiler, Alyssa. "9 Surprising Reasons Why You Should Smile More." *Inspiyr*. August 2, 2014.

Episode 025 System Theories, Racism, and Human Relationships
T.K. Coleman, education director at Praxis
tkcoleman.com
Bertalanffy, Ludwig von. *General System Theory: Foundations, Development, Applications*. 1969.

Episode 026 Mindfulness is Complexity's Best Friend
Pilar Angel, co-founder of Axeos Performance Institute

Episode 027 Applying Design Thinking to Complex, Wicked Problems
Angie Cross (see "About the Authors")
Stacy Hale (see Episode 000)
Design4Emergence: A Magazine for the Network Age. design4emergence.com
Rogers, K. H., R. Luton, H. Biggs, R. Biggs, S. Blignaut, A. G. Choles, C. G. Palmer, and P. Tangwe. "Fostering Complexity Thinking in Action Research for Change in Social-Ecological Systems." *Ecology and Society*. 2013
Creativity at Work. creativityatwork.com

Episode 028 Leaping into Design Thinking
Turi McKinley, thought leader and executive director at frog design
frogdesign.com
frogdesign.com/services/camp
McKinley, Turi. "Design Thinking: Lead Change in your Organization." *Lynda.com*. 2016.

Episode 029 Design-Led Innovation
Jon Kolko, vice president of design at Blackboard and founder of the Austin Center for Design
jonkolko.com
Kolko, Jon. "Design Thinking Comes of Age." *Harvard Business Review.* September 2015.
BiggerPictureVideo. "Learning Graphic Facilitation—7 Elements by Bigger Picture." *YouTube.* 2013.
Kolko, Jon. *Well-Designed: How to Use Empathy to Create Products People Love.* 2014.

Episode 030 How to Apply Human-Centered Design
Andrea Ippolito, lead at the Department of Veteran Affairs Innovator's Network
MITsdm. "How to Build a Culture of Innovation through Design and Systems Thinking." *YouTube.* 2015
Department of Veterans Affairs Center for Innovation. "Designing for Veterans: A Toolkit for Human-Centered Design." 2015.
va.gov/playbook/downloads/vaci-project-toolkit.pdf

Episode 031 Understanding Complexity in Context
Diego Espinosa (see Episode 006)
T.K. Coleman (see Episode 025)
Gloria Burgess (see Episode 024)
Jason Dykstra (see Episode 003)
Hanson, Barbara G. *General Systems Theory: Beginning with Wholes.* 1995.

Episode 032 Exploring Complex Human Networks
Mary Anne Herrick, communications program officer at World Vision and president of Foster Care Alumni of America (FCAA)
fostercarealumni.org
FCAA's 2018 fund raising campaign. crowdrise.com/fcaa
World Vision. "Water is Life | World Vision." *YouTube.* 2016.

Episode 033 Human Stories for Network Change
Jasper Faolan, psychiatric nurse practitioner and founder of Journal to Save Your Life
journaltosaveyourlife.org
Groogan, Greg. "Selfie Guy Defeats Depression with Motherload of Goodwill." *FOX 26 News Houston.* May 26, 2016.

Episode 034 Network Challenges Within Child Welfare
D'Artagnan Caliman, chief of staff for the Maryland Department of Human Resources and founder of Caliman and Associates
"Breaking the School-to-Prison Pipeline for Young Offenders One Class at a Time." *PBS News Hour.* June 29, 2016.
Liptak, Adam. "U.S. Prison Population Dwarfs that of Other Nations." *The New York Times.* April 23, 2008.

Episode 035 Using Systems Science to Improve Community Health
Daniel Taber, food policy and systems scientist
dantaberphd.com
Johns Hopkins University. "Systems Science and Obesity." coursera.org

Episode 036 Harnessing System Potential
Angela Montgomery, author and co-founder of Intelligent Management
intelligentmanagement.ws
Montgomery, Angela. *The Human Constraint*. 2016.
socialnation.it/en
Intelligent Management. "Building the Core Conflict Cloud."
intelligentmanagement.ws/learningcentre/what-are-the-thinking-process-tools
Goldratt, Eliyahu M. and Jeff Cox. *The Goal: A Process of Ongoing Improvement*. 2016.
Goldratt, Eliyahu M. *The Theory of Constraints*. 1987.

Episode 037 The Complexity of Living Organizations
Oleg Konovalov, strategy and change management consultant
Konovalov, Oleg. *Organisational Anatomy: A Manager's Guide to a Healthy Organisation*. 2016.

Episode 038 A Philosophical Look at Company Culture
Benjamin Taylor, managing partner at RedQuadrant and chief executive at The Public Service Transformation Academy
redquadrant.com
publicservicetransformation.org
SCiO, a not-for-profit group for systems practitioners. scio.org.uk
LinkedIn group, "Systems Thinking Network."
linkedin.com/groups/2639211
The Outliers Inn podcast. theoutliersinn.com

Episode 039 Company Culture as an Emergent and Evolving Phenomenon
Sonja Blignaut, founder of More Beyond
morebeyond.co.za
Snowden, J. Dave and Mary E. Boone. "A Leader's Framework for Decision Making." *Harvard Business Review*. November 2007.
Blignaut, Sonja. "Entrepreneuring Meaning." *TEDxPretoria*. July 8, 2016.

Episode 040 The Myth of the Dysfunctional System
Benjamin Taylor (see Episode 038)
Angela Montgomery (see Episode 036)
Oleg Konovalov (see Episode 037)
Sonja Blignaut (see Episode 039)
Montgomery, Angela. "Dysfunctional Organization: Definition and Cure." *Intelligent Management*. August 3, 2016.
Weisbord, Marvin. "Techniques to Match to Our Values." *marvinweisbord.com*. 2005.

Episode 041 Compassion in Chaos
John Feehery, Republican strategist and lobbyist
thefeeherytheory.com

Episode 042 The Neuroscience Behind Your Vote
Spencer Gerrol, founder and chief executive officer of SPARK Neuro
sparkneuro.com
Bogage, Jacob. "Whom Are You Voting For? This Guy Can Read Your Mind." *The Washington Post*. June 23, 2016.

Episode 043 Conscious Voting Through Political Curiosity
John Feehery (see Episode 041)
Spencer Gerrol (see Episode 042)
I Side With..., an interactive political website. isidewith.com
The Other Side, a documentary. theothersidedocumentary.com
"The Story of Citizens United v. FEC." *The Story of Stuff Project*.

Episode 044 Making Sense of Human Systems
Dave Snowden, founder and chief scientific officer of Cognitive Edge
cognitive-edge.com
Snowden, J. D. "Multi-Ontology Sense Making: A New Simplicity in Decision Making." *Cognitive Edge*. May 1, 2005.
Snowden, J. Dave and Mary E. Boone. "A Leader's Framework for Decision Making." *Harvard Business Review*. November 2007.

Episode 045 Human Networks Inspire Holiday Cheer
Angie Cross (see "About the Authors")
Haley Campbell-Gross (see "About the Authors")
Shoemaker, Paul. *Can't Not Do: The Compelling Social Drive That Changes Our World*. 2015.

Episode 046 A Sneak Peek at Our Leadership Series
Scott Jancy, writer and growth consultant
scottjancy.com
Paolo Gaudiano, president and chief technology officer at Icosystem
icosystem.com

Episode 047 Visionary Leadership in a Complex World
Scott Jancy (see Episode 046)

Episode 048 The Intersection of Diversity, Innovation, and Performance
Paolo Gaudiano (see Episode 046)
Gaudiano, Paolo and Ellen Hunt. "The Complexity of Diversity." *Forbes*. July 11, 2016.

Episode 049 Using Data and Design to Empower Teams
Rico Andrade, vice president of marketing at iCharts
icharts.net
McChrystal, Stanley A. *Team of Teams: New Rules of Engagement for a Complex World*. 2015.
Catmull, Edwin E., and Amy Wallace. *Creativity, Inc.: Overcoming the Unseen Forces That Stand in the Way of True Inspiration*. 2014.
Kelly, Kevin. *Out of Control: The New Biology of Machines, Social Systems, and the Economic World*. 1994.

Episode 050 Complexity Science Is Everyone's Science
Yaneer Bar-Yam, researcher, complexity scientist, and founding president of the New England Complex Systems Institute (NECSI)
necsi.edu/yaneer-bar-yam
Bar-Yam, Yaneer. *Making Things Work: Solving Complex Problems in a Complex World*. 2004.

Episode 051 Learning from Nature
Laurie Marshall, author and education transformer
laurie-marshall.com
Beating the Odds Now! for Educators. beatingtheoddsnow.com
Glaser, Judith. *Conversational Intelligence*. 2014.

Episode 052 Developing Innovative Ecological Systems
Chris Ling, director of the School of Environment and Sustainability at Royal Roads University
researchgate.net/profile/Christopher_Ling
royalroads.ca/people/chris-ling

Episode 053 Becoming Environmentally Conscious in a Complex World
Laurie Marshall (see Episode 051)
Chris Ling (see Episode 052)
James Greyson, founder and chief executive officer of BlindSpot Think Tank
blindspot.org.uk
Greyson, James. "Seven Policy Switches for Global Security." *NATO Science for Peace and Security Series C: Environmental Security*. 2010.
The Story of Stuff. storyofstuff.org
Earth Day. earthday.org
Bioneers, an environmental not-for-profit organization. bioneers.org

Episode 054 Reversing Complex Problems for Global Security
James Greyson (see Episode 053)
Greyson, James. "Let's Unshrink Thinking and Reverse Reverse-Progress." *TEDxBradford*. December 30, 2014.
James Greyson. "Climate Rescue Can Cookstove Instructions." 2050kids.org/an-inconspicuous-hero

Episode 055 Unintended Consequences
Yaneer Bar-Yam (see Episode 050)

Episode 056 The Science of How Nature Affects Our Brain
Florence Williams, science journalist and author
florencewilliams.com
Williams, Florence. *The Nature Fix: Why Nature Makes Us Happier, Healthier, and More Creative*. 2018.

Episode 057 A Curious Conversation on the Brain and Body
Angie Cross (see "About the Authors")
Haley Campbell-Gross (see "About the Authors")
Sapolsky, Robert. "The Biology of Our Best and Worst Selves." *TED Talks*. April 2017.

Episode 058 Unlearn Toxic Habits and Retrain Your Brain
Eileen McDonald, co-founder and chief coaching officer of the Axeos Performance Institute

Episode 059 A Closer Look at the Female Brain
Louann Brizendine, neuropsychiatrist and founder of the Women's and Teen Girls' Mood and Hormone Clinic
about.me/louannbrizendine
Brock, Sharon. "Neuropsychiatrist Explains Why Female Brains Are Different." *UCSF*. April 9, 2007.
Brizendine, Louann. *The Female Brain*. 2009.
Brizendine, Louann. *The Male Brain: A Breakthrough Understanding of How Men and Boys Think*. 2010.
Brizendine, Louann. "Louann Brizendine at TEDxBerkeley." *TEDxBerkeley*. May 13, 2013.
ONE Media. "The Female Brain Official Trailer (2018) Sofía Vergara, James Marsden Comedy Movie HD." *YouTube*. January 9, 2018.

Episode 060 Exploring the Science of What Connects Us
Cassandra Vieten, president of the Institute of Noetic Sciences (IONS)
noetic.org/profile/cassandra-vieten

Episode 061 Finding Purpose Through Connection
Laurie Marshall (see Episode 051)
Andrea Dennis, director of development, outreach, visibility, and engagement (DOVE) at Greenheart International
greenheart.org
Jon Darrall-Rew, European director of the Global Purpose Movement
globalpurposemovement.com/the-purpose-summit-berlin
IONS 18th International Conference. noetic.org/conference
Webster, de'Ante. *The Flood of Kindness: Inspired by Hurricane Katrina*. 2017.

Episode 062 Mysteries of the Extended Mind
Rupert Sheldrake, author and researcher in the field of parapsychology
sheldrake.org
Sheldrake, Rupert. *Science Set Free: 10 Paths to New Discovery*. 2012.
Sheldrake, Rupert. *Dogs That Know When Their Owners Are Coming Home: And Other Unexplained Powers of Animals*. 2011.
Sheldrake, Rupert. *Science and Spiritual Practices: Reconnecting Through Direct Experience*. 2017.

Episode 063 A Philosophical Look at the Mind
Robert Burton, neurologist and author
rburton.com
Burton, Robert A. *A Skeptic's Guide to the Mind: What Neuroscience Can and Cannot Tell Us About Ourselves*. 2014.
Burton, Robert A. *On Being Certain: Believing You Are Right Even When You're Not*. 2009.
Burton, Robert A. "Our World Outsmarts Us." *Aeon*. May 3, 2017.

Episode 064 Our Evolving and Complex Human Brain
Rick Hanson, psychologist and senior fellow of the Greater Good Science Center at UC Berkeley
greatergood.berkeley.edu/profile/rick_hanson
rickhanson.net

Episode 065 Hardwiring Happiness
Rick Hanson (see Episode 064)
The Foundations of Well-Being, program by Rick Hanson.
thefoundationsofwellbeing.com
Hanson, Rick. "Hardwiring Happiness." *TEDxMarin 2013*. November 7, 2013
Hanson, Rick. *Hardwiring Happiness: The New Brain Science of Contentment, Calm, and Confidence*. 2016.

Episode 066 Conference on Complex Systems 2017
Carlos Gershenson, editor-in-chief of *Complexity Digest* and a co-chair of the Conference on Complex Systems
unam.academia.edu/CarlosGershenson
Gershenson, Carlos. *Complexity: 5 Questions*. 2010.

Episode 067 The Path to Economic Growth According to Complexity Science
Yaneer Bar-Yam (see Episode 050)
Bar-Yam, Yaneer, et al. "Preliminary Steps Toward a Universal Economic Dynamics for Monetary and Fiscal Policy." *New England Complex System Institute*. December 29, 2017.

Episode 068 Managing for Happiness with Complex Systems Science
Jurgen Appelo, speaker and entrepreneur
jurgenappelo.com
Appelo, Jurgen. *Managing for Happiness: Games, Tools and Practices to Motivate Any Team*. 2016.
Appelo, Jurgen. *How to Change the World: Change Management 3.0*. 2012.
Brace, Louise. "12 Steps to Worker Happiness and Beyond." *Management 3.0*. July 2, 2015.
Happy Melly, a global professional happiness association. happymelly.com
Appelo, Jurgen. "Managing for Happiness." *TEDxLille*. July 21, 2016.
Human Current. "Jurgen Appelo's 12 Steps to Happiness." *YouTube*. October 26, 2017.

Episode 069 Designing Meaningful Work in Complex Times
Bud Caddell, founder of NOBL Collective
budcaddell.com
nobl.io
Caddell, Bud. "How Complex Systems Will Save Us." *TEDxIndianapolis*. November 19, 2014.

Episode 070 Living from the Inside Out
Sam Mednick, journalist and podcast host
sammednick.com/bio
Blueprint Coaching. blueprintcoaching.ca
Happy Melly's podcast. happymelly.com/our-story

Episode 071 Throwback to 2017
Yaneer Bar-Yam (see Episode 050)

Episode 072 Feeling Complexity with Radical Wholeness
Philip Shepherd, author and developer of The Embodied Present Process
philipshepherd.com
Shepherd, Philip. *Radical Wholeness: The Embodied Present and the Ordinary Grace of Being*. 2017.
Shepherd, Philip. *New Self, New World: Recovering Our Senses in the Twenty-First Century*. 2010.

Episode 073 Coherence and the Heart's Intuitive Guidance
Rollin McCraty, psychophysiologist and executive vice president and director of research at HeartMath Institute
heartmath.org/about-us/request-a-speaker/rollin-mccraty
heartmath.org/research/global-coherence
Human Current. "An Explanation of Coherence with HeartMath's Dr. Rollin McCraty." *YouTube*. December 2, 2018.

Episode 074 The Systems Worldview of Human History
Jeremy Lent, author and founder of the nonprofit Liology Institute
jeremylent.com
Lent, Jeremy R. *The Patterning Instinct: A Cultural History of Humanity's Search for Meaning.* 2017.
Lent, Jeremy. "The Patterning Instinct." *YouTube*. April 4, 2017.

Episode 075 Embracing Complexity
Jean Boulton, author and visiting fellow with the Complex Systems Research Centre at Cranfield School of Management
Embracing Complexity, Jean Boulton and Claremont Management Consulting Ltd website. embracingcomplexity.com
Boulton, Jean G., Peter M. Allen, and Cliff Bowman. *Embracing Complexity: Strategic Perspectives for an Age of Turbulence*. 2015.
Human Current. "Jean Boulton—How Complex Systems Science Informs Climate Change." *YouTube*. April 19, 2018.

Episode 076 The WorkHuman Movement
Mike Wood, *WorkHuman Radio* host and social marketing manager at Globoforce
workhuman.com

Episode 077 Exploring Artificial Intelligence
Melanie Mitchell, professor of computer science at Portland State University and external professor at the Santa Fe Institute
Mitchell, Melanie. *Complexity: A Guided Tour*. 2011.
Human Current. "AI & Machine Learning with Melanie Mitchell." *YouTube*. December 4, 2018.

Episode 078 Loving AI
Julia Mossbridge, lead robot psychologist at Hanson Robotics, director of the Innovation Lab at the Institute of Noetic Sciences, and founder of the Mossbridge Institute
Barušs, Imants, and Julia Mossbridge. *Transcendent Mind: Rethinking the Science of Consciousness*. 2017.
Science and Nonduality. "Not just a Robot—I am Sophia." *YouTube*. January 23, 2018.

Episode 079 The Integration of Human and Artificial Intelligence
Alfredo Morales, assistant professor at the New England Complex Systems Institute and visiting scientist at the Massachusetts Institute of Technology Media Lab
ajmoralesguzman.com
media.mit.edu/people/alfredom/overview

Episode 080 Bring Your Human to Work
Erica Keswin, author and founder of the Spaghetti Project
ericakeswin.com/the-spaghetti-project
Keswin, Erica. *Bring Your Human to Work: 10 Sure-Fire Ways to Design a Workplace That Is Good for People, Great for Business, and Just Might Change the World*. 2019.

Episode 081 The Science, History, and Implications of Real Magic
Dean Radin, author and chief scientist at the Institute of Noetic Sciences (IONS)
deanradin.org
noetic.org
Radin, Dean I. *Real Magic: Ancient Wisdom, Modern Science, and a Guide to the Secret Power of the Universe*. 2018.
Three Bridges West. "Real Magic." *YouTube*. November 27, 2017.

Episode 082 The Complexity and Chaos of Creativity
Jon Kolko (see Episode 029)
Kolko, Jon. *Creative Clarity: A Practical Guide for Bringing Creative Thinking into Your Company*. 2017.

Episode 083 Relationship Building and System Innovation
Amber Christian, speaker and the founder of ConsultAce
consultace.biz/about

Episode 084 Recapping WorkHuman
Jennifer Fondrevey, chief humanity officer
jenniferjfondrevay.com
Angela Romei, director at Panorama
panoramaglobal.org

Episode 085 How Networks Learn
César Hidalgo physicist, complexity scientist, and professor at Massachusetts Institute of Technology
chidalgo.com
media.mit.edu/groups/collective-learning/overview
Datawheel, a company that helps make big data accessible. datawheel.us
Hidalgo, César. *Why Information Grows: The Evolution of Order, from Atoms to Economies*. 2016.

Episode 086 The Conscious Evolution of Leadership
Renee Moorefield, co-founder and chief executive officer of Wisdom Works
wisdom-works.com
Powered by Wisdom Works, *Be Well Lead Well* helps individuals and organizations thrive. bewellleadwell.com
Be Well Lead Well assessment for leaders. bewellleadwellpulse.com
Senge, Peter M. *The Fifth Discipline: The Art and Practice of the Learning Organization*. 2006.
Wheatley, Margaret J. *Leadership and the New Science: Discovering Order in a Chaotic World*. 2011.
Wilber, Ken. *A Theory of Everything: An Integral Vision for Business, Politics, Science, and Spirituality*. 2001.
Gallwey, W T. *The Inner Game of Work: Focus, Learning, Pleasure, and Mobility in the Workplace*. 2001.
Anderson, Robert J., and William A. Adams. *Mastering Leadership: An Integrated Framework for Breakthrough Performance and Extraordinary Business Results*. 2015.

Episode 087 Holistic Science
Jean Boulton (see Episode 075)
Postgraduate courses in Holistic Science at Schumacher College, schumachercollege.org.uk

Episode 088 The Science and Philosophy of Complexity
Carlos Gershenson (see Episode 066)
Carlos Gershenson's website. turing.iimas.unam.mx/~cgg
Self-organizing Systems Lab. turing.iimas.unam.mx/sos
Gershenson, Carlos. "Bringing Urban Technology To Life With Realtime Feedback." *YouTube*. June 28, 2012

Episode 089 Nonlinear Systems and Technology with Complexity Labs
Joss Colchester, founder of Complexity Labs (now Systems Innovation)
systemsinnovation.io
Systems Academy channel on *YouTube*
Systems Academy. "Blockchain: An Overview." *Udemy.com*

Episode 090 Reclaiming Leadership for the Human Spirit
Margaret Wheatley, doctor of education and founder and president of The Berkana Institute
margaretwheatley.com
berkana.org
Wheatley, Margaret J. *Who Do We Choose to Be?: Facing Reality, Claiming Leadership, Restoring Sanity*. 2017.
Wheatley, Margaret J. *Leadership and the New Science: Discovering Order in a Chaotic World*. 2011.
Warrior Training 2019.
margaretwheatley.com/2019-warriors-for-the-human-spirit-training-new
Stuart Kauffman is a complex systems researcher.
en.wikipedia.org/wiki/Stuart_Kauffman
Fritjof Capra. fritjofcapra.net
Cape Cod Institute. cape.org

Episode 091 Localism and the Economics of Happiness
Helena Norberg-Hodge, filmmaker and founder of Local Futures
localfutures.org
Norberg-Hodge, Helena. *Ancient Futures*. 2016.
The Economics of Happiness. "Going Local: The Solution-Multiplier." *Vimeo*. 2017.

Episode 092 ICCS 2018
Yaneer Bar-Yam (see Episode 050)

Episode 093 From Venezuela to a Science of Collapse
Ricardo Hausmann, professor of economic development at Harvard University's Kennedy School of Government and director of the Center for International Development at Harvard University
ricardohausmann.com
International Conference on Complex Systems 2018. necsi.edu/iccs-2018
Columns by Ricardo Hausman.
project-syndicate.org/columnist/ricardo-hausmann
The Observatory of Economic Complexity. atlas.media.mit.edu
Hausmann, Ricardo, César A. Hidalgo, Sebastián Bustos, Michele Coscia, Alexander Simoes, and Muhammed A. Yıldırım. *The Atlas of Economic Complexity: Mapping Paths to Prosperity*. 2014.

Episode 094 A Complex Systems Approach to Countering Mind Control
Steven Hassan, founding director of the Freedom of Mind Resource Center
freedomofmind.com
Program in Psychiatry and the Law. pipatl.org
Freedom of Mind Resource Center. freedomfromundueinfluence.org

Episode 095 Answering Hard Questions
Ryan Singer, head of strategy at Basecamp
basecamp.com
Ryan Singer's website. feltpresence.com

Episode 096 Macroeconomics and Financial Crises
Carmen Reinhart, professor at Harvard University's Kennedy School of Government
carmenreinhart.com

Episode 097 Applied Mathematics and the Evolution of Music
Natalia Komarova, professor at the University of California, Irvine
math.uci.edu/~komarova

Episode 098 School Reform with Complexity Theory
Patrick McQuillan, professor at Boston College
bc.edu/bc-web/schools/lynch-school/faculty-research/faculty-directory/patrick-mcquillan.html

Episode 099 Mitigating Risk in Biotechnology
George Church, professor of genetics at Harvard Medical School and professor of health science and technology at Massachusetts Institute of Technology
arep.med.harvard.edu/gmc
The BRAIN Initiative. braininitiative.nih.gov
Church, George M., and Ed Regis. *Regenesis: How Synthetic Biology Will Reinvent Nature and Ourselves*. 2012.

Episode 100 Cheers to 100 Episodes
Richard Cooper, board member of the New England Complex Systems Institute and professor of international economics at Harvard University
scholar.harvard.edu/cooper
Emma Towlson, professor of physics at Northeastern University and researcher in the Center for Complex Network Research
emmatowlson.github.io
The Center for Complex Network Research, directed by Albert-László Barabási.
barabasilab.com/people/emma-towlson
Emma Towlson's Antarctic voyage.
chuffed.org/project/women-science-antarctica-towlson
Homeward Bound. homewardboundprojects.com.au

Episode 101 A Journey of Computational Complexity
Stephen Wolfram, creator of Mathematica, Wolfram|Alpha, and the Wolfram Language, and the founder and chief executive officer of Wolfram Research
stephenwolfram.com/about
wolfram.com/mathematica
wolframalpha.com
wolfram.com/language
Wolfram, Stephen. *A New Kind of Science*. 2002.

Episode 102 Probability and Adaptation
Gad Saad, evolutionary behavioral scientist and marketing professor at Concordia University, Montreal, Canada
concordia.ca/jmsb/faculty/gad-saad
Gad Saad channel on *YouTube*
Saad, Gad. *The Evolutionary Bases of Consumption*. 2007.
Nassim Nicholas Taleb, probability researcher and former trader
Nassim Nicholas Taleb's website. fooledbyrandomness.com
Taleb, Nassim N. *Skin in the Game: Hidden Asymmetries in Daily Life*. 2018.
Taleb, Nassim N. *The Black Swan: The Impact of the Highly Improbable*. 2010.

Episode 103 Evolutionary Theory and Behavioral Science
Gad Saad (see Episode 102)
Human Current. "Evolutionary Behavioral Science with Professor Gad Saad." *YouTube*. September 11, 2018.
Saad, Gad. *The Consuming Instinct: What Juicy Burgers, Ferraris, Pornography, and Gift Giving Reveal About Human Nature*. 2011.

Episode 104 Systems-Based Precision Medicine
Daniel Stickler, co-founder and chief medical officer of The Apeiron Center for Human Potential
apeironcenter.com/our-team
apeironzoh.com

Episode 105 Modeling and Forecasting Infectious Diseases
Irene Bosch, research scientist at Massachusetts Institute of Technology
Elena Naumova, mathematician and director of the Tufts University Initiative for Forecasting and Modeling of Infectious Diseases
nutrition.tufts.edu/profile/faculty/elena-n-naumova

Episode 106 Networks and Interdisciplinary Research
Jose Mateos, physicist and network researcher at the Universidad Nacional Autónoma de México
unam.mx
cssociety.org

Episode 107 Stories from Network Science PhDs
Brennan Klein, a network science PhD student at Northeastern University
jkbrennan.com
Brennan Klien's art website. jkrofling.com
Sarah Shugars, a network science PhD student at Northeastern University
sarahshugars.com

Episode 108 An Exploration of Social Minds
Simon DeDeo, assistant professor in social and decision sciences at Carnegie Mellon University and external professor at the Santa Fe Institute
cmu.edu/dietrich/sds/people/faculty/simon-dedeo.html
santafe.edu/people/profile/simon-dedeo
Laboratory for Social Minds. tuvalu.santafe.edu/~simon
DeDeo, Simon. "Play, Where Ideas Come From." *TEDxIndianaUniversity*. December 20, 2016.

Episode 109 The Social Pursuit of Order
Loren Demerath, author and professor of sociology at Centenary College of Louisiana
lorendemerath.com
centenary.edu/academics/departments-schools/sociology
Demerath, Loren. *Explaining Culture: The Social Pursuit of Subjective Order*. 2013.

Episode 110 A Holistic Perspective of Evolution
Josh Mitteldorf, author and theoretical biologist
joshmitteldorf.scienceblog.com
Mitteldorf, Josh, and Dorion Sagan. *Cracking the Aging Code: The NewScience of Growing Old—And What It Means for Staying Young*. 2016.
How to Live a Long Time. agingadvice.org

Episode 111 Stories and Insights from the Complexity Community
Rosa Benito, physics professor at the Universidad Politécnica de Madrid
Roozbeh Daneshvar, blockchain software engineer at International Business Machines (IBM)
Victor Dossetti, professor at Benemerita Universidad Autonoma de Puebla
Javier Borondo, research scientist and chief technology officer of AGrowingData
agrowingdata.com

Episode 112 Gaia Theory, Holistic Science, and Deep Ecology
Stephan Harding, founding member of Schumacher College
schumachercollege.org.uk/about/stephan-harding
Harding, Stephen. *Animate Earth: Science, Intuition and Gaia*. 2009.
Deep Time Walk. deeptimewalk.org
James Lovelock, developer of Gaia hypothesis. jameslovelock.org
Human Current. "Expanding the Western Worldview with Dr. Stephan Harding." *YouTube.* November 7, 2018.
Harding, Stephen. "Gaia Theory and Deep Ecology." *YouTube*. July 8, 2013.

Episode 113 Connecting with Complexity Thinkers
Brad Glisson, co-founder of Thoughtblox
bradglisson.com
thoughtblox.com
Liz Johnson, managing editor of the *Journal of Policy and Complex Systems*
policyandcomplexsystems.wordpress.com
Vinesh Raja, bioinformatician and software engineer
Rhys Lindmark, podcaster and head of community and long-term societal impact at Massachusetts Institute of Technology Media Lab

Episode 114 Your Vote Brings Democratic Stability
Yaneer Bar-Yam (see Episode 050)
Alexander Siegenfeld, PhD physics student at Massachusetts Institute of Technology
Siegenfeld, Alexander and Yaneer Bar-Yam. "Negative Representation and Instability in Democratic Elections." *New England Complex Systems Institute*. October 29, 2018.
Human Current. "Complexity Science Research on Democratic Elections." *YouTube*. November 2, 2018.

Episode 115 The Network Science of Success
Albert-László Barabási, author and professor of network science at Northeastern University
barabasi.com
khoury.northeastern.edu/people/albert-laszlo-barabasi
Barabási, Albert-László. *The Formula: The Universal Laws of Success*. 2018.
Barabási, Albert-László. *Linked: How Everything Is Connected to Everything Else and What It Means for Business, Science, and Everyday Life*. 2014.
Barabási, Albert-László. *Bursts: The Hidden Patterns Behind Everything We Do, from Your E-Mail to Bloody Crusades*. 2011.

Episode 116 An Educator's Guide to Systems Thinking
Linda Booth Sweeney, systems educator and author
lindaboothsweeney.net
Sweeney, Linda B., and Dennis L. Meadows. *The Systems Thinking Playbook: Exercises to Stretch and Build Learning and Systems Thinking Capabilities*. 2013.
Sweeney, Linda B., and Jana Christy. *When the Wind Blows*. 2015.
Meadows, Dennis L, Linda B. Sweeney, and Gillian Martin-Mehers. *The Climate Change Playbook: 22 Systems Thinking Games for More Effective Communication About Climate Change*. 2016.
Sweeney, Linda B. *Connected Wisdom: Living Stories About Living Systems*. 2009.
Sweeney, Linda B. *When a Butterfly Sneezes: A Guide for Helping Kids Explore Interconnections in Our World Through Favorite Stories*. 2001.
Sweeney, Linda B., and Jana Christy. *When the Snow Falls*. 2017.
World Wildlife Fund event, "Living Systems," presented in collaboration with National Geographic. worldwildlife.org/pages/living-systems#
Sweeney, Linda B. "How Children's Stories Help Us See Real-World Patterns." *Medium*. March 16, 2018.

Episode 117 Influencing Pattern-Based Change
Denise Easton, author and co-founder of Complexity Space Consulting
complexityspace.com
"The Commons" is a Plexus Institute project. plexusinstitute.org/our-work/the-commons-project
Easton, Denise, and Lawrence Solow. *Complexity Works!: Influencing Pattern-Based Change in Teams and Organizations*. 2016.
Taylor, Mark C. *The Moment of Complexity: Emerging Network Culture*. 2009.
Brown, Dan. *Origin: A Novel*. 2018.
SAM, a new political party. joinsam.org

Episode 118 Practical Applications for Interconnected, Nonlinear Systems
Tanuja Prasad, entrepreneur in the social impact sector
linkedin.com/in/tanujaprasad

Episode 119 Navigating Systemic Risk and Critical Transitions
Len Fisher, author and biophysical scientist
Fisher, Len. *Crashes, Crises, and Calamities: How We Can Use Science to Read the Early-Warning Signs*. 2011.
lenfisherscience.com
Fisher, Len. *Rock, Paper, Scissors: Game Theory in Everyday Life*. 2010.
Fisher, Len. *Scientific Discovery from the Brilliant to the Bizarre: The Doctor Who Weighed the Soul, and Other True Tales*. 2013.
Fisher, Len. *How to Dunk a Doughnut: The Science of Everyday Life*. 2003.

Episode 120 The Social Impact of Intelligent Systems
Mihaela Ulieru, founder and president of the IMPACT Institute for the Digital Economy
theimpactinstitute.org
weforum.org/agenda/authors/mihaela-ulieru
Endor, AI-powered business predictions. endor.com
Sapien, a social network platform. sapien.network

Episode 121 Preparing Leaders for Complex Change
Deborah Westphal, chief executive officer of Toffler Associates
tofflerassociates.com/about/leadership/deborah-westphal
Toffler Associates. *Leadership for Human-Centric Organizations: Creating Resiliency in the Midst of Disruption*. 2018.
Toffler, Alvin. *Future Shock*. 1994.
Toffler, Alvin. *The Third Wave*. 1980.

Episode 122 Design-Led System Transformation
Daniel Wahl, whole systems design practitioner and educator
danielchristianwahl.com
Daniel Wahl's *Medium* blogsite. medium.com/@designforsustainability
Meadows, Donella. "Dancing with Systems." *The Donella Meadows Project Academy for Systems Change.*
Project Drawdown. drawdown.org
Gaia Education. gaiaeducation.org/face-to-face/trainers/daniel-christian-wahl
Wahl, Daniel C. *Designing Regenerative Cultures*. 2016.

Episode 123 Complexity Theory and Systemic Risk in Financial Markets
Francesco Filia, chief executive officer of Fasanara Capital
fasanara.com
Filla, Francesco. "Risk Assets Downturn is Starting to Play Out, Analyst Says." *CNBC*. November 14, 2018.
Filla, Francesco. "Look for Nonlinearities in the Market, Expert Says." *CNBC*. September 14, 2018.

Episode 124 Health and Aging as Complex Networks
Alan Cohen, researcher and professor at Université de Sherbrooke, and director of Cohen Labs
alancohen.recherche.usherbrooke.ca/en/director
cdrv.csss-iugs.ca/researcher-in-details/43-cohen-alan

Episode 125 Learning How to Thrive in Complexity
Jennifer Garvey Berger, researcher and coach at Growth Edge Coaching
cultivatingleadership.co.nz/our-team/jennifer-garvey-berger
growthedgecoaching.com
Berger, Jennifer G. *Unlocking Leadership Mindtraps: How to Thrive in Complexity*. 2019.

About the Authors

Angela (Angie) Cross has a natural ability to see the big picture of interconnectedness and how ecosystems work together, all while also zooming in to appreciate specific parts. It's no wonder that the saying often attributed to Aristotle, "The whole is greater than the sum of its parts," is one of her favorites.

Cross, who was born in Canada and moved to the United States in her mid-twenties, feels deeply connected to the Earth (Gaia) and is passionate about human equality. She has twenty-five years of leadership and engagement experience. When Cross first met Douglas (Doug) Drane she was finishing up her bachelor of arts degree in organizational communication at St. Edward's University—a degree that developed her systems thinking lens. Cross would spend hours with Doug while he discussed his complexity model and all the ways it could be applied to business and ultimately used to make the world a better place. It was Doug, who soon became a valued mentor to Cross, who expanded her mind into the world of complexity thinking. Doug inspired and encouraged Cross to go back to school where she earned her master of arts in leadership from Royal Roads University in Canada—a degree plan that had a strong focus on leading with a complexity thinking lens. While earning her MA, she co-founded and began co-hosting *Human Current*.

Cross is incredibly grateful for the opportunities she had to learn from Doug, establish *Human Current*, interview amazing-brilliant guests, and develop a deep friendship with her co-host, Haley. These experiences have made a lasting impression in her life, helped shape who she is, and prepared her for her next chapter.

Haley Campbell-Gross is an open-minded explorer, content marketer, writer, and systems thinker who is curious about the systems and patterns that shape our world. She is a deep thinker, questioner, and truth-seeker and has always felt at odds with the "us versus them" narrative found in Western culture. Campbell-Gross continues to believe that seeing the world in circles rather than straight lines (to paraphrase Peter Senge) is the key to bringing humanity closer to the truth of who we are, why we are here, and how we can make a difference.

Campbell-Gross earned her bachelor's degree in marketing from the University of Houston and has a passion for growing communities around a brand. When she met her mentor Douglas (Doug) Drane, she was a new graduate looking to make a difference in the world as a marketer. Doug inspired her to spread the word about complexity science and grow a community of systems thinkers and change agents. Haley joined the *Human Current* team and quickly fell in love with the community of guests and listeners that emerged from the podcast. She is very grateful to Doug and her wonderful co-host and friend, Angie, for the opportunity to learn and grow through this magnificent project. Learning about complexity science has forever changed her life and she is excited to continue to grow, both personally and professionally, as a systems thinker and change agent.

www.ingramcontent.com/pod-product-compliance
Lightning Source LLC
LaVergne TN
LVHW050918080826
845145LV00001B/121

* 9 7 8 0 5 7 8 5 0 9 8 3 9 *